October 18, 2007

AF697221

Marion Cowdell
12810 W. Blue Bonnet Dr.
Sun City West,
Arizona 85375

Dear Marion,

Thank you for your order for 1 (one) copy of the book "Copperfield Remembered". It will be shipped today.

Even though I grew up in Copperfield I learned quite a lot about the earlier days of the town and of the Utah Copper mine while preparing this book. I hope you enjoy reading it as much as I enjoyed writing it.

Yours truly,

R Eldon Bray

R. Eldon Bray

Bingham Canyon.net
c/o Eldon Bray
2186 McKinley Drive
Grand Junction, CO 81503
Tel. 970-245-8209

Marion,
Cost of the two books and shipping is $27.95. Thanks if you've already mailed it.

The book "Copperfield Remembered" was printed August 19, 2007. It contains more than 142 photos, with descriptions, of the town and of the Utah Copper dating from before 1900 to 2003. It also includes interesting stories of the townspeople and of the mine. It can be bought per the following Purchase Order.

PURCHASE ORDER

Books can be purchased over the internet at BinghamCanyon.net or by mail from:

BinghamCanyon.net
c/o Eldon Bray
2186 McKinley Drive
Grand Junction, CO 81503
Tel. 970-245-8209

Date ____

Name ____

Street ____

City ____

Zip ____________ **Phone #** ____________________

PAYMENT BY CHECK OR MONEY ORDER PLEASE

Signature ____

ITEM WANTED		QUANTITY	COST	TOTAL
Book "COPPERFIELD REMEMBERED"				
One or two copies	$24.95 each	________	X $24.95	________
Three to five copies	$21.95 each	________	X $21.95	________
Six to ten copies	$19.95 each	________	X $19.95	_________
Sub Total				________
S&H $3/Order				$3.00
TOTAL ORDER		# ______		$________

Figure 1 – Copperfield and Utah Copper Mine Looking NW About 1938

From the Gordon Bodily Collection - This photo shows most of the features of Copperfield that many of us remember. The E-Line Bridge is on the right edge of the photo; it was built in 1912 and demolished about 1941. The main street of Copperfield extends up the canyon from the bridge to the lower left of the photo. The U.S. Hotel and the ore bins, dump, and smokestacks of the Niagara Tunnel are halfway up Main Street. The Circle is a short distance up the street from the hotel. Some of the Company houses along Main Street are evident. The Upper Terrace two-story duplexes are on the right of the photo; they were built in 1925-1927. An edge of Dinkeyville is in the lower right corner of the photo. A train of loaded ore cars is coming down the hill toward Fox's Switch that is in the lower left of the photo. The Niagara Tunnel, with stacks of lumber, is to the left of the loaded ore cars. Jap Camp is the group of long single-story wooden buildings on the left of the photo above Fox's Switch. Greek Camp is above the Niagara Tunnel this side of the open-pit mine. The edge of the pit has been expanded to very near the E-Line Bridge and is approaching Greek Camp. Mine dumps are extending far down the canyon on the West side. Telegraph and the U.S. aren't shown in this photo; they would be to the left. The location of the Bingham-Copperfield vehicular tunnel can't be seen but it could be present and would help to more closely date this photo. A guesstimate is that this photo was taken between 1937 and 1940.

ABOUT THE FRONT COVER

Close-up of 1926 Black-and-White Photo of Copperfield Used for Front Cover - With Permission of Utah State Historical Society, all rights reserved, from the Shipler Collection. The front cover, composed of a 1926 photo and a 2003 photo, shows that the present-day Utah Copper open-pit mine swallowed the old town of Copperfield. This 1926 photo shows many interesting features. Two steam engines can be seen – one on H-Level in Dinkeyville and one in the Fox's Switch tailtrack by the Copper Water. Four wagons are being pulled by teams of horses up the steep grade above the Copper Water. The pipeline that carried compressed air from the compressor behind the U.S. Hotel to the mines in Galena Gulch (The U.S.'s Jordan and Galena mines and maybe to the Silver Shield and other mines) is above the Copper Water. There are some tanks attached to the pipeline just below the concrete overpass. The pipeline goes through the overpass, up the road past the H-Level tailtrack and then straight up the hill (the pipeline shortcut) toward the U.S.. The concrete abutments above Fox's tailtrack, to the right of the Copper Water, were probably for the Upper Crossing (the Utah Copper's H-Level Wooden Bridge). The partial wooden trestle just above the last houses on the east side of Main Street could have been for the Copper Belt Railroad line that went up Galena Gulch – it predated all of the Utah Copper operations including the waste dumps that filled the lower part of Galena Gulch. The name of the mine under the Jap Camp road is unknown. The addition to the rear of the schoolhouse is under construction.

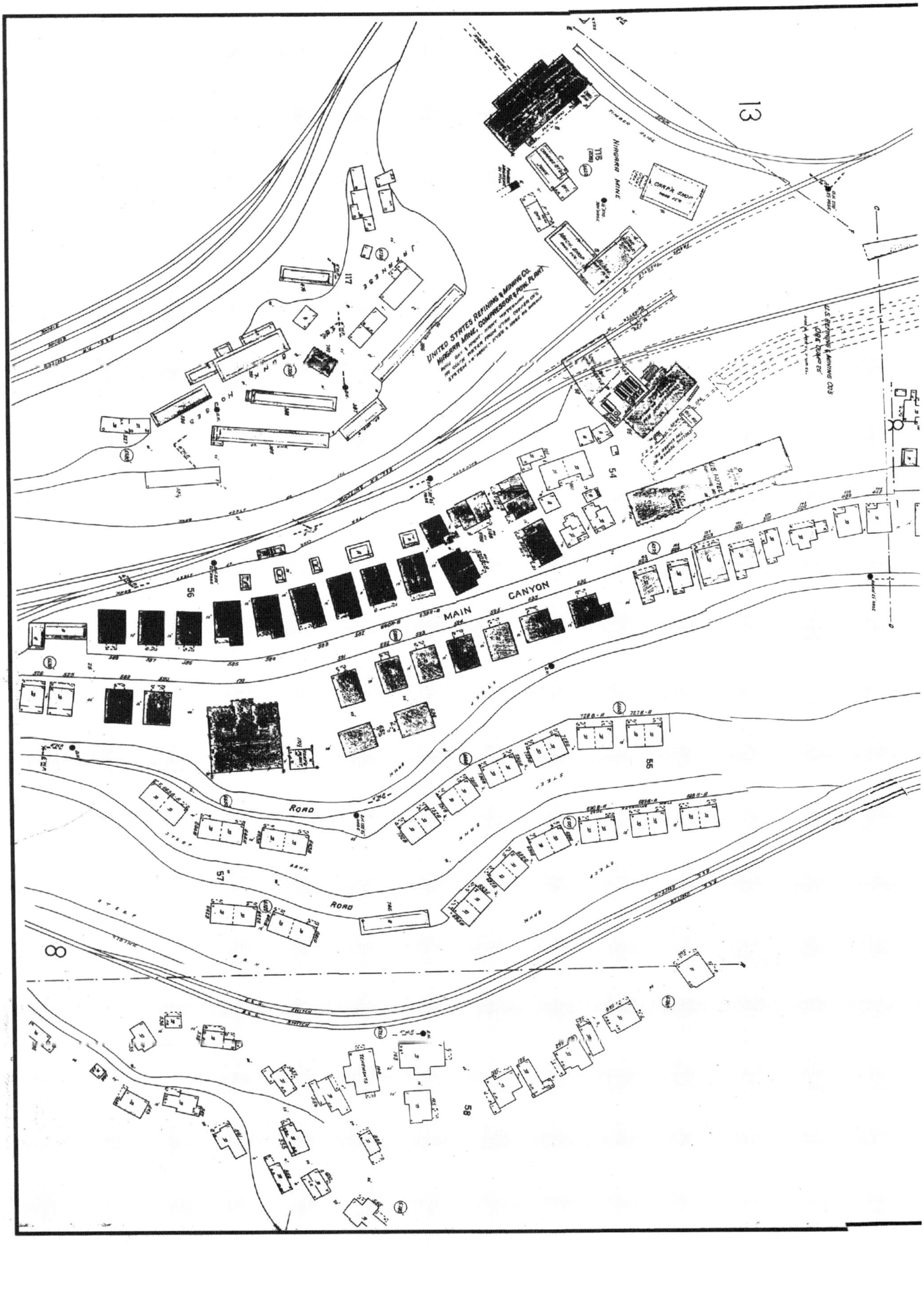

13
NIAGARA MINE
115
UNITED STATES REFINING & MINING CO.
NIAGARA MINE, COMPRESSOR & POW. PLANT
117
JAPANESE
BUNK HOUSES
54
56
MAIN CANYON
55
ROAD
57
ROAD
8
58

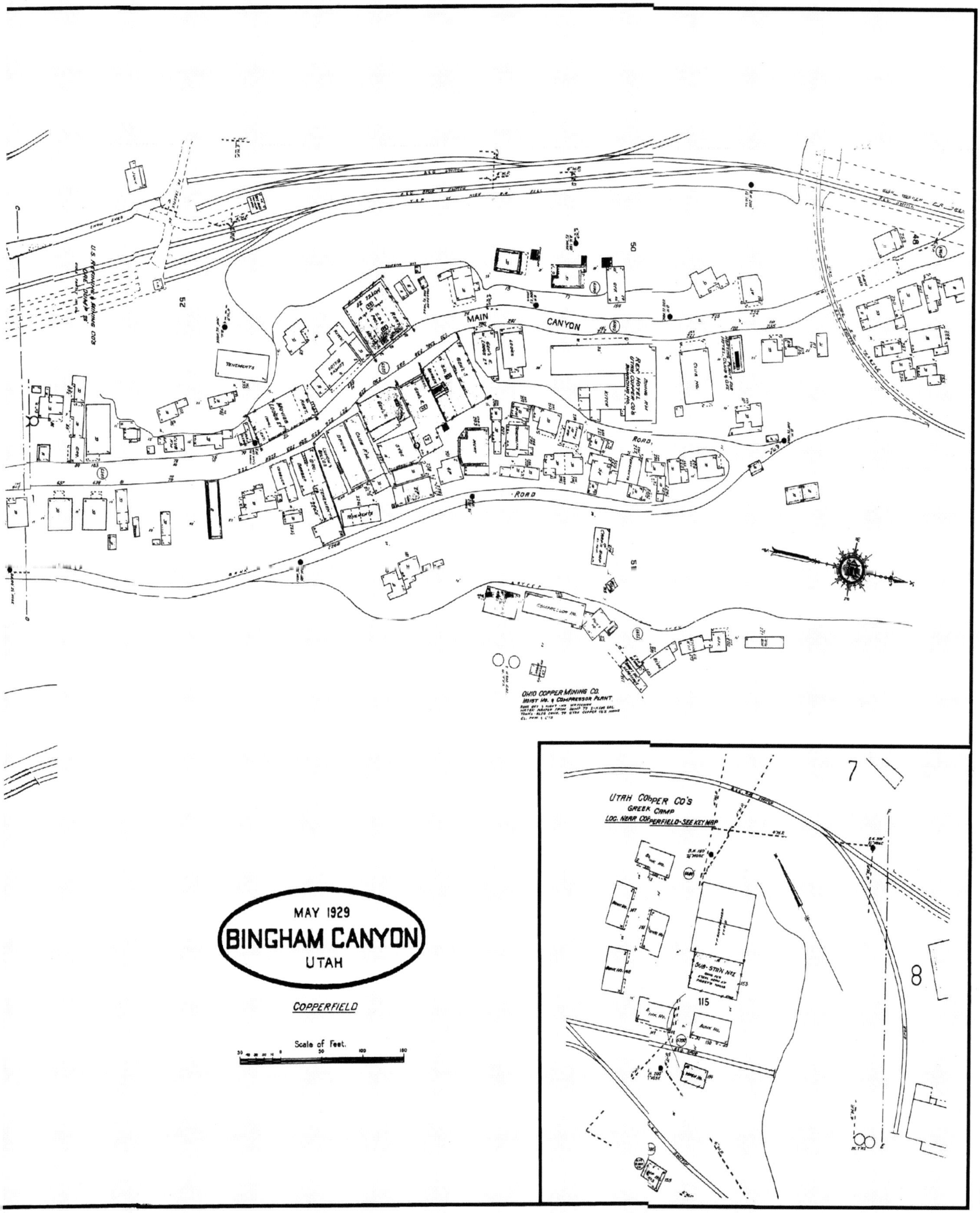
MAY 1929
BINGHAM CANYON
UTAH
COPPERFIELD
Scale of Feet.
MAIN CANYON ROAD
ROAD
TENEMENTS
CLUB HO.
OHIO COPPER MINING CO.
HOIST HO. & COMPRESSOR PLANT
UTAH COPPER CO'S
GREEK CAMP
LOC. NEAR COPPERFIELD-SEE KEY MAP
SUB-STA'N NO.1
115
7
8
52
51
50
48

From "Down Memory Lane, **My First Job**" by John J. Creedon
16 Mar 1921(excerpts) Courtesy of Gordon Bodily

One of the most outstanding man made Bingham landmarks, now all but completely removed or covered over, was the old Tramway skirting the left side of the Canyon up to the mouth of Bear Gulch, where it crossed the Canyon on a wooden trestle and continued on the right wall of Jordan and Galena Gulch.

This tramway was built in the days of hand work and black powder, in the late "80's" by the Old Jordan Company. The track was 39 inch gauge, the same as the old narrow gauge railroad, originally built into Bingham by the D&RGW railroad, later made over to standard gauge width.

This tramway was taken over by Standish and Jimpson, two early day business men, and operated in connection with their coal and lumber business, which was later sold to George E. Chandler, who incorporated the Bingham Coal and Lumber Company.

My first remembrance of the tramway was as a small boy. I watched Grandpa Tibble (grandfather of Clyde) work on track repairs. I remember his old white horse that lived to an old age. This tram way was operated to convey supplies up to the Old Jordan Mine and cars loaded with ore and dropped down the steep grade by hand brakes.

On the trip up to the mine horses were used, two or three in tandem according to the size of the load on each car and the trip usually consisted of ten or twelve cars requiring some 20 to 25 horses.

Here is where I came in, for I was hired to lead the string of horses back to the yard, my first job. I rode the lead horse followed by the others with the halter strap of one tied to the tail of the preceding horse. Not many people now living, can recall seeing a small boy of 11 years, about 65 pounds, riding the first horse with the others trailing along down the three miles of dusty road in the summer and the mud, knee deep in the early spring.

There was no sidewalks in the canyon and anyone that was unlucky enough to be along side of the muddy road enjoyed the pleasure of a mud bath.

When I was asked by Mr. Jimpson one day, to drive a one horse passenger car to take Mr. and Mrs. Ray Bemis from the Bingham train to the Galena Mine, I was on my own. There was no means of informing the ore train coming down that the passenger car was on the way up-each had to watch for the other. On this day, the horse, "Ned" was more on the job than myself, because just at the sharp turn below Colonel Wall's Starless Mine, I saw Ned's ears stand up and almost at the same instant, Ned jumped down hill off the track and the ore train hit the passenger car, throwing us to the ground. Mrs. Bemis remained in the car.

The ore train came to a stop with each car touching the one ahead. However, the passenger car was not thrown off the track and started down grade with Mrs. Bemis on the floor between the two lengthwise seats, with no one in control.

One of the ore train drivers was my good friend, Victor Finnel, who ran along side and put on the brake and stopped the car with minor bruises to the three of us. The men lifted the passenger car off the track, so the ore train could pass, then lifted it back on the track and "Ned" and I proceeded to complete the trip. I braked the car slowly back to the yard and "Ned" came home on his own power.

The men who made up the ore train were the leader, L.B. "Ren" Jenkins, Dave Wherret, Joe Shelley, Oscar Christopherson, Victor Finnel, George Hobbs and Gene Barton, all fine men.

The preceding article came from the pen of my good friend, Isador Gauchet, who I am happy to report has recovered from a severe heart attack suffered in January.

COPPERFIELD

REMEMBERED

A History of Copperfield and Adventures While Growing Up There

By R. ELDON BRAY

BinghamCanyon.net 2007

Library of Congress Catalog Card No.

ISBN 978-1-4243-0858-3

PRINTED IN THE UNITED STATES OF AMERICA

BY F. C. PRINTING, SALT LAKE CITY, UTAH

TABLE OF CONTENTS

ILLUSTRATIONS

Questions regarding places, dates, and happenings have been inserted throughout this book. If you have answers or further questions please feel free to address them to R. Eldon Bray, 2186 McKinley Drive, Grand Junction, CO 81503. Information regarding especially pertinent photos that are not generally available will be welcome

FOREWORD

Copperfield was one of several small communities in Bingham Canyon, Utah. In its early days it was called "Upper Bingham". Nestled in the upper reaches of Bingham Canyon on the east side of the Oquirrh Mountains twenty miles from Salt Lake City, Bingham became famous as the home of the world's largest open-pit copper mine.

Copperfield was, for about twenty years while I was growing up there, the site of the observation point for the huge open-pit mine and was thus exposed to the view of millions of tourists from all over the world. Most of these tourists were taken aback as they maneuvered up the narrow canyon with streets so confined that when traffic was heavy the movement of automobiles was restricted to one-way with only inches to spare on each side. And then they drove through the mile and a quarter long Bingham-Copperfield vehicular tunnel that was far longer and more claustrophobic than most of them had ever experienced. Finally the tourists emerged from the tunnel into the light of day in Copperfield. They were again subjected to the narrow streets and were set upon by young boys and girls who offered to sell them souvenirs or to act as guides to show them the open-pit copper mine. After these trials the tourists finally arrived at the observation point and beheld the Utah Copper open-pit mine. The mine was (and is now even more so) a truly astonishing wonder!

Within the enclosed upper reaches of the canyon, immediately adjacent to the enormous open pit mine, the people of Copperfield led lives that they didn't realize were so unique. They accepted the narrow streets, the long tunnel, the huge open-pit mine and numerous other unusual places, objects of interest, and activities as parts of their everyday lives. Copperfield was, indeed, a community with a unique history and lifestyle. The last of the town was mined away in 1958 and it is only now, almost five decades later, that I have come to realize how much it meant to me. Copperfield can never be replaced – it can only live on in our memories.

This book was originally intended to tell the personal experiences of me and my family as I grew up in Copperfield. However, several compelling stories were recalled that involved some of my friends so these stories were also included. And then there were interesting incidents on the job and in other parts of Bingham so they were included. While discussing one "happening", many people would recall another that was equally interesting. So the book kept expanding until my personal history was only a small part of it.

The biggest expansion came about with the gathering of a "few" photographs so that the reader could "see" the town as it was from the mid 1930's to its demise in the late 1950's. The "few" photos grew to over a hundred. At that point it was apparent that a description of the photos would amount to a history of the town going clear back to the 1860s. After all the photographs were in their "final" place another would be found that just had to be included. Many of these photos are of the open-pit mine, not only of Copperfield; I hope you enjoy them.

The final book is a conglomeration of the picture-history of the town of Copperfield and the surrounding area and of the Utah Copper mine, especially entertaining human interest stories that include other people and other parts of town and my personal history and family adventures. My family and personal history, which were originally intended to make up the whole book, have been relegated to the end.

It is apparent that over the hundred years or so of Copperfield's existence there are many more wonderful stories and historical facts that could and should be written about Bingham. Unfortunately most of the old-timers have passed away and a wealth of information has been lost with them. I would like to gather as many photos, as much historical information and as many human interest stories as are available as soon as possible for inclusion in one or more additional books. These additional books might include "On the Job" stories of Utah Copper or Kennecott as well as the U.S. and Lark mines, and "Adventure and Humor in Bingham" that would include stories from all parts of Bingham, Copperton, and Lark. For example, a couple of stories that remain to be written are those of the stick of "dynamite" that was tossed into the shack full of men who fought their way out through the windows and the trackshifter operator who insisted, contrary to directions, on going up the switchback and ended up leaping from the shifter just before the train of loads ran into him.

You are invited to submit stories for inclusion in the next book. Submissions would have to include your permission for publication. They would be subject to review and editing and to my determination of their suitability. Wherever practical you will be given credit for your contributions.

DEDICATION

This book is dedicated to my Mom and Dad and to my hunting buddies who include George Kubota and Warren Palmer who are no longer with us and to my brother Garland and to Lee Halverson. My (or our) dog Tip was an indispensable partner. We walked many miles together over the hills surrounding Copperfield, shot thousands of .22 shells, and brought home hundreds of cottontails and a goodly amount of other game. The hills were more alive when we were there.

ACKNOWLEDGEMENTS

Thanks are given especially to Eugene Halverson, Salvador Aguayo, Isabell Scroggin, and to my brother Garland for their contributions and their review of the material in this book. Many other individuals, most of them former residents of the Copperfield and Bingham area, have contributed information that is included herein. Information about a single incident was often gathered from several sources so that credit for contributions may not be given to any one individual.

Organizations from which photographs were obtained include Kennecott Utah Copper, the Utah State Historical Society, University of Utah, U.S. Geological Survey and Bingham High School Alumni.

Individuals who provided photos include Lori (Clinton) Beckstrand, Gordon Bodily, Eugene Halverson, Max (Barbara) Ivie, Johnny (Sandra) King and Isabell (Shirl) Scroggin. Gordon Bodily's collection of photographs on CDs was especially useful because it was so nearly complete and was in digital format (thanks to Craig Bodily); Gordon claims no ownership of the photos but he put forth a lot of effort in gathering them and making them available. The original sources of many of the photographs are difficult or nearly impossible to identify because they have been widely copied and distributed over the years. However, many of Gordon's photos came from Kennecott Utah Copper and Dee Johanson. Scott Crump helped in organizing Gordon's photo collection; copies of Gordon's collection were then given to

Scott who donated those from Dee Johanson to the Utah State Historical Society where they now reside as The Dee Johanson Collection, C1536; others, especially from Kennecott were donated to the Bingham High School Alumni collection. Several credits are given to "Bingham, The Greatest Copper Camp In The World" by The Bingham Commercial Club, 1909" because the photos, as far as can be determined, first appeared there; however, most of these photos are now available at the Utah State Historical Society. Special thanks to Vera Johnson who was an avid photographer; she took hundreds of photos of the U.S. – Copperfield area but passed away years ago. Vera's photos are available through the efforts of her daughter Velma and of Willimay McDonald Tervort, Isabell Scroggin and Eugene Halverson who generously shared Vera's photos with others including myself.

Much of the early history of Copperfield and Bingham was obtained from Professional Paper 38 of the U.S. Geological Survey. Writings of John Creedon provided additional information.

Human interest stories were substantiated whenever practical by discussions with those who were personally involved. In some instances the recollections of several individuals were melded into one final account. Bud Patrick helped to substantiate and was also involved in several of these stories.

Special thanks to Bruce Manchee, Instructor at the Western Colorado Community College, and his students for design of the cover for this book.

ENDORSEMENT

Relating his own experiences as well as the stories he heard as a boy growing up in Copperfield, Eldon Bray provides readers with a rich, entertaining, and insightful glimpse into life in Utah's copper towns during the 1940s and 1950s. From selling copper ore to tourists to a BB gun shootout on a Copperfield road, Bray's autobiographical history of his long gone hometown will amuse and inform readers of all ages.

Allan Kent Powell
Historian
Utah State Historical Society

BRIEF HISTORY AND INTRODUCTION

Copperfield, or Upper Bingham, came into being shortly after the first mineral claim was staked in Bingham Canyon, Utah in 1863. The town grew up around and supported the early lead-silver and gold underground mines and the placer gold creek-bed mines. The high costs of transportation and supplies caused the underground mines to be uneconomic in the 1860s. However, placer gold in the creek beds could be recovered and immediately sold or used to buy other goods – this helped the town to survive during its early years. After the railroad was connected to Salt Lake City in 1870 the costs of transportation and goods were reduced and for the next thirty years underground mining was generally profitable. Copper was at first an unprofitable hindrance to the mining and processing of the lead-silver and gold ores. Near the turn of the century, however, copper became more valuable and interest turned to exploiting the huge low-grade deposit on "The Copper Hill". In 1906 the Boston Con and Utah Copper open-pit mines commenced operation. Copperfield was on the very edge of the Utah Copper open-pit mine which became the largest open-pit copper mine in the world. The town was constantly threatened and nibbled at by the mine as it expanded. Copperfield endured on the very edge of the open-pit until 1958 when the last vestiges of the town were finally demolished and devoured by the mine.

Most of us as we grew up in Copperfield knew very little of the history of the town and of the features that we saw there every day. In 1944, when I was ten years old, the town already had a history of almost eighty years. Many mines had been started and shut down; many people had come and gone or died. There were relicts of the past all around us. The "Shady Track" that we walked along just above Main Street, the "Cocoa Dirt" that we played with above H-Tailtrack, Fox's Switch, the many old mine dumps that spotted the hillsides -- we took them for granted and never heard or read of their history because it wasn't immediately available to us. We didn't know who to talk to or where to find any written history. We didn't even know what questions to ask. The existence of the huge tree stumps above the Silver Shield mine were common knowledge but again, we never learned their history or took any photos of them.

It has been a delight for me to learn more of the history of Copperfield. For example, the "Shady Track" was actually the "Shay Track" of the Copper Belt Railroad along which small Shay steam engines pulled ore cars. The soft brown clay-like "Cocoa Dirt" was the finely-ground tailings of a mill that operated near Telegraph. Fox's Switch was named after the owner (Mrs. Fox) of an adjacent boarding house that was demolished to make way for company-built houses. And the stumps of the huge trees above the Silver Shield Mine were part of "The Big Grove" that were used to build the roof of the Mormon Tabernacle in Salt Lake City.

The history of Copperfield includes the story of its demise. From the first day that open-cut (later referred to as open-pit as the mine grew deeper) mining began in 1906 the town was threatened and inexorably chewed away by the Utah Copper open-pit mine. Finally, in 1958, in one final, last big gulp, the entire town was demolished and the last remnants were mined away. Now, as we view the open-pit mine from the Visitor's Center, there is nothing to be seen of the Copperfield that we grew up in – it lives on only in our memories.

This book is an attempt to preserve in photo and writing the town of Copperfield and some of the events that transpired there. It includes a photo history and description of Copperfield from its earliest days to its last. Interesting incidents are recounted that involved one-time residents of the town and of workers at the Utah Copper mine. The last part of the book tells of the author and his family's personal adventures in Copperfield.

More than 140 photos, dating back to the 1890s, show many features of the town that had been erased or of which only vestiges remained by the 1930s. The Holden Tramway, Copper Belt Railroad, the aerial tramway from the U.S. mines in Galena Gulch, the Upper Crossing on H-Level, and the Big Grove in Silver Shield were unfamiliar, forgotten features. The E-Line Bridge, U.S. Noodle House, Dianna Theater and Miner's Merc were still remembered by some older former residents but not by the younger. One feature that practically everybody remembers is the enormous cottonwood tree in Bear Gulch that we called "Big Tree"; unfortunately the author couldn't find a photo of it.

The stories that are related brought many a chuckle as they were recounted. Two reactions to this book have been "If kids did that today they'd be put in jail!" and "If a mining company did that today they'd be sued!" Needless to say, life was very different in those days and Copperfield was a very unique little town.

The author, R. Eldon Bray, lived on the Terrace Heights section of Copperfield from the age of one year until after he graduated from Bingham High School in 1952. He worked for Kennecott for over twenty years, eighteen of them as a Geologist.

MINING MAN REPORTS ON PARK-BINGHAM

I.F. Block, prominent mining engineer of Salt Lake City, visited the properties of the Park-Bingham Mining Company in Butterfield on Monday last and has made the following report:

The Park-Bingham property in Bingham, through which the United States Smelting, Refining and Mining Company is driving the Butterfield tunnel to connect with the lower workings of the United States Mine, is no longer a prospect – it is a mine. There are a number of fissures which can be drifted on a distance of approximately 150 feet or raised on about 125 feet, and the limestone entered. Furthermore pay ore can be followed the entire distance to the limestone, where, without doubt, large beds of ore will be opened up.

"In my entire experience, I have never seen more favorable conditions for development.

The formation first entered consisted of big stocks of monzonite and diorite in which are intercalated small beds of limestone. A network of fissures cuts the whole block of ground. These fissures range from a few inches to 12 feet in width.

"The first fissure of any prominence, called fissure No. 1, has a strike of due magnetic north with a dip varying from 45 to 65 degrees. This fissure, measuring from four to five feet wide, traverses a limestone bed about seven feet thick included between diorite walls. The ore is of excellent grade. Some of it carries as high as 65 per cent lead, $3.20 in gold and 457 ounces in silver. Bunches and streaks of pyrargyrite show throughout the whole mass, which should average not less than $60 a ton."

The bed of quartzite next entered by the tunnel, explained Mr. Block, was 115 feet wide. This was cut by fully twenty parallel fissures ranging from 6 inches to twelve feet in width, not a single one of which but carried pay ore. The last 100 feet of the Butterfield tunnel, now about on the line separating Park-Bingham from United States ground, is now in a soluble Sappington limestone.

"The last fissure," said Mr. Block, "exposed in the limestone ninety feet from the Silver Shield line, carried from fifteen to eighteen inches of solid shipping ore carrying thirty-seven ounces of silver, $1.80 in gold, 2 per cent copper and 15 per cent lead.

"All of these fissures offer ideal conditions for prospecting. The calcite showing in the vein matter proves conclusively that the limestone lies above or it could not have been carried there by downward percolating water. In the last fissure the streaks of pyrargyrite which show all through the vein matter carry as high as 1000 ozs. of silver to the ton.

"The strength of the mineralizing solutions must have been very intense, else the quartzite and diorite, generally believed not to be favorable to replacement, would not be ore-bearing as they are. Likewise, the fissures show immense strength or they would not persist to this horizon from 1600 to 1700 feet below the surface."

The above article was printed in the March 22, 1924 edition of The Press-Bulletin, Bingham Canyon, Utah.

WHERE WAS MY HOUSE?

Now just where was it – the place where I grew up? Not only was my old house gone but the whole neighborhood and even the entire town were no longer there. I felt like a homeless vagabond with no place to call his own.

The view from the lip of the huge open-pit copper mine offered few clues to the previous existence of the Terrace Heights, of Copperfield or, indeed, any part of the town of Bingham Canyon with its miles of streets and hundreds of homes. They were all gone -- either mined out or covered over with waste dumps by the Utah Copper mine. The site of my old home must now be occupied by a space up in the air hundreds of feet above one of the mine levels along which huge haulage trucks now bustled.

MY HOME TOWN

By Salvador (Chinky) Aguayo

How I envy all of you. You can go back whenever you have a mind to and visit the town that you grew up in. You can go back to the place of your roots and visit the old schoolhouse you learned to read and write in. You can look at that old building which may be old and dilapidated but it is the one where the memories will come rushing back to you just by standing there and looking. Your memories can go back to the good teachers you had and maybe to some bad ones. You will remember the friends you had while going to this school – the first girl you had a crush on or maybe the first girl that had a crush on you. You can walk around the streets of the town and remember the stores that used to be there before the new buildings were put in their place after the old ones were demolished. Just standing there in the street you can remember the parades on the 4th of July, the fireworks, and the races with the quarters and dimes as prizes.

Yes, I envy all of you that can go back to your home town and sharpen your memories of days gone by because I have only my memories to reflect on. The town I spent my youth in is gone. There is no remnant of the town to sharpen my mind – nothing to focus on and bring into sharper remembrance those long-gone days. No brick is left of the old school house where I studied for six years. The very street I walked on is no more. No old-timer sits on his front porch chatting with anyone passing by about the old days. There is no porch, there is no house, and the old-timer has moved twenty or a thousand miles away.

The copper mine that used to be the cause for our existence in the city had growing pains and swallowed everything within its grasp. Homes, buildings, streets, and all the signs of vibrant life that had kept the mine operating are gone. All that is left is in the minds of thousands of others like myself that remember the wonderful days of living in the Melting Pot of Races and Nationalities of all description. The cultures of dozens of peoples merging into an army of workers intent on extracting a living – the way of life – from Mother Earth.

Every few years I go back to that giant hole in the ground just so I can guess as to where the town used to be but when I do I envy all of you even more. All I can see with my eyes is a big scar on the earth. Only when I close my eyes and think back to those carefree days of my youth can I remember Bingham Canyon as it was.

CHAPTER ONE

HISTORY OF COPPERFIELD

Copperfield was part of the town of Bingham Canyon which is located in the Oquirrh Mountains about twenty miles southwest of Salt Lake City. When the Mormon Pioneers came to the Salt Lake Valley in 1847 Bingham Canyon was a heavily forested wilderness. Red pine (Douglas fir) trees were plentiful so the canyon became a source of timber for the pioneers. The canyon was also used for livestock grazing.

Bingham Canyon was thrust into prominence in 1863 by the discovery of mineral and the location of the earliest mining location in Utah. Copperfield (also known as Upper Bingham) was an important part of Bingham. The central part of the town which included the E-Line Bridge, the U.S. Hotel and the Copperfield Grade School was a mile and a half above Carr Fork and half a mile below the intersection of Galena and Bear Gulches. Copperfield was especially prominent in its early years when underground mines in the immediate area were first being developed and areas of population were located adjacent to the mines and mills.

Placer gold was found in the gravels of Bingham Creek in 1864, a year after the first discovery of mineral. Gravel washing for placer gold was actively pursued from the spring of 1865 with maximum production from 1868 to 1872 after which it declined. Some of this placer gold mining occurred in the Copperfield area.

Lode (underground) mining, after an initial burst of activity in 1864, was essentially put on hold until 1870 because the costs of transportation and of goods were too high. Placer mining helped to maintain the economy and activity in all of Bingham Canyon during the lull in lode mining. The railroad was connected to Salt Lake City in 1870 after which lode mining activity was markedly renewed and continued to accelerate. New underground mines were brought into production throughout Bingham Canyon.

After the initial discoveries of lode mineral and placer gold hundreds of men came to Bingham to hunt for metals and to work in the mines. In the mid and late 1860s people were already living and working in the area that became known as Copperfield. From 1870 to 1906 underground mining, first for lead and silver with some gold and later (after 1896) for copper also, was the dominant activity. In 1891 and 1892 there were 21 producing underground mines in the Bingham area including several in or near Copperfield.

Open-cut copper mining was started by the Boston Consolidated Mining Company in 1906 and a few months later by the Utah Copper Company. The open cut (the two companies and their mines merged) quickly became the largest producer in Bingham Canyon. It soon dominated the town as "The

Hill" (later referred to as "The Pit" expanded in the narrow canyon. At first tentacles of the mine, in the form of railroad tracks that connected the mine to the waste dumps, were extended behind and above the town. Then overburden nearer and nearer to Copperfield was removed. Next the open-pit mine chewed away the roadway and buildings between the main part of Bingham and Copperfield. Finally it swallowed the entire town of Copperfield as well as most of Highland Boy and Carr Fork and some of the Main Street of Bingham. In a last burst of activity essentially all remaining reminders of the several communities that once occupied Bingham Canyon were covered over by waste dumps. Now (2007) the open pit continues to expand but its progress is no longer marked by the engulfment of any homes, schoolhouses or businesses -- they were all obliterated decades ago. The open-pit mine will continue to grow still deeper and its slopes may become steeper for a few more years but ultimately it will no longer be profitable. Then, perhaps, the "roots" of the open pit will be mined by underground methods.

The long-term future of the Utah Copper mine is dependent on the profitability of mining the deep, high-grade copper orebody that lies to the north beneath the old town of Bingham and Carr Fork. It has been announced that underground mining may soon commence on this large, high-grade mineralization. It will be expensive to mine but the values in copper, gold, and silver are anticipated to more than offset the high costs. If it can be mined at a profit then the Utah Copper mine should remain in operation for decades into the future.

MRS. KOLTSKI NOT TO BLAME

Her Testimony Showed That the Dog Had Deliberately Disregarded Orders She Gave Him.

Judge Oscar Hallam, in this address before the South Dakota Bar association told this good one:

"I recall a case of assault and battery, against the Koltski family, and in the course of the examination it developed that the Koltski family dog had taken an active part. Mrs. Koltski, when on the stand, was asked if she didn't instigate the activities of the dog. She insisted that she did not. The attorney said: "Didn't you say, "Sic 'em, Caesar?" for that was the dog's name. She said, "No, I did not."

"The attorney said, 'You said something to the dog?'

"She replied: 'Well, what if I did?'

"He insisted: 'Tell us what you said to the dog.'

"She answered: 'Why, I said, "Don't sic 'em, Caesar."'"

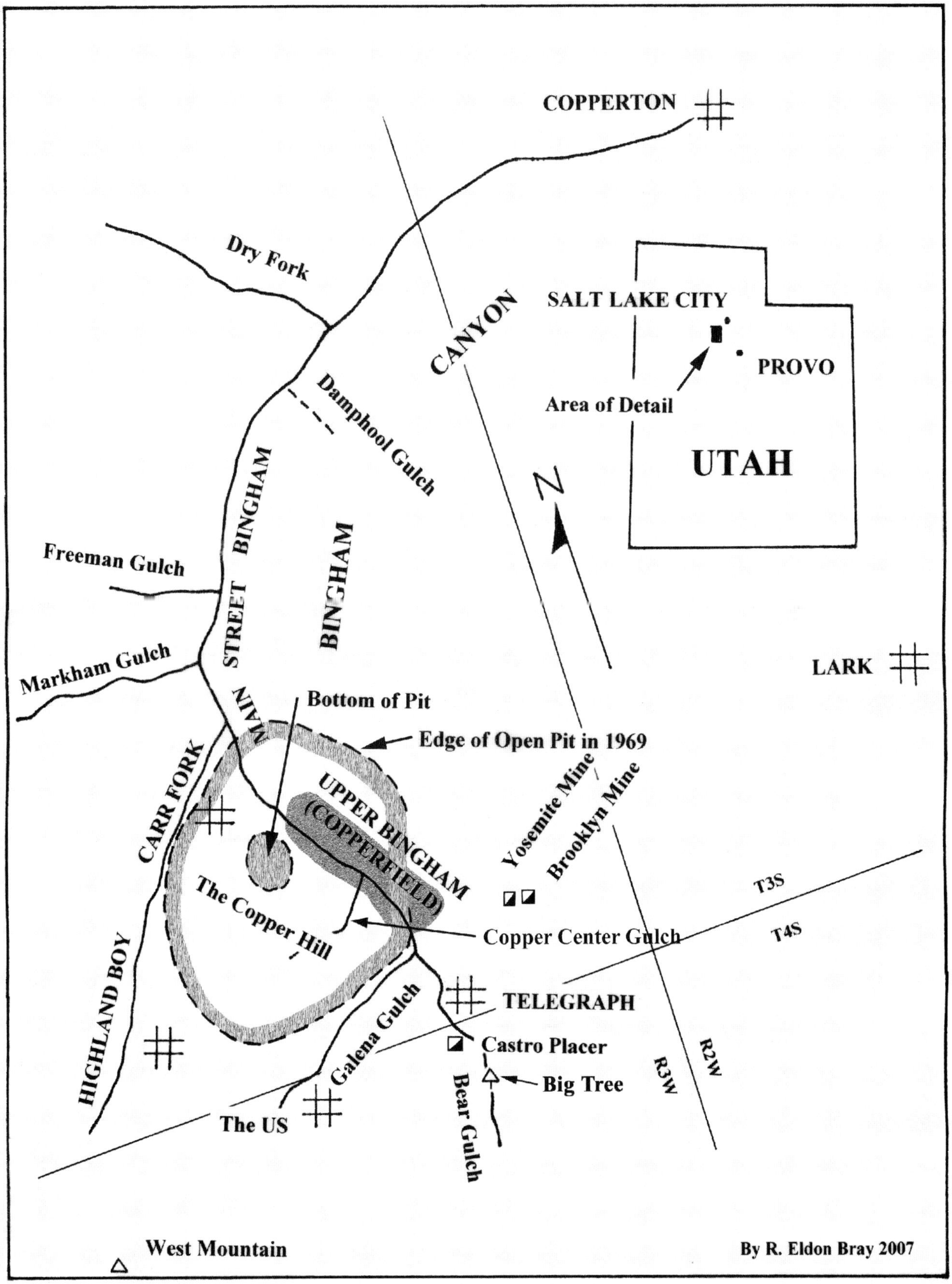

Figure 1– Location Map of Bingham Canyon, Copperfield and Immediate Area

Drawn by R. Eldon Bray 2007 – The outline of the open-pit mine is superimposed over the roads and other features that predated the open pit. The Copper Hill and the town of Copperfield were mined away by the Utah Copper open-pit mine. Some minor features are shown because they are mentioned or have photos elsewhere in this book. The largest gold nugget was found near the mouth of Damphool Gulch. There are photos of the Brooklyn and Yosemite Mine and of the Castro Placer. Big Tree is shown because it was so well known to residents of Copperfield for many decades. West Mountain (Sunshine Peak) stood at the head of Bingham Canyon and was familiar to most residents of the area.

CAR LOAD OF WHISKEY FOUND IN BINGHAM AUTO

Deputy Sheriff, While Searching for Automobile That Killed a Salt Lake Woman, Stumbles on to Nearly $5000 Worth of Liquor Hid in Jack Creighton's Car – The Auto Was Found in the Bingham Garage, and It Is Believed that Officers Will Find the Notorious Bootlegger Who Has Been Bringing Large Quantities of Liquor to Bingham – Deputy Sheriff V. B. Jones Makes the Discovery.

Last Tuesday noon the chief deputy sheriff of Bingham Canyon received a message from the sheriff's office in Salt Lake to apprehend any automobile driver who possessed a car the right fender of which was bent. This order was an effort on the part of the Salt Lake County authorities to capture the driver who ruthlessly ran over and killed a lady on South State street Monday night, as the car which figured in the accident suffered a bent right fender from the collision.

Chief Deputy Oscar Fullmer and his force immediately began a search of the camp for a car with a bent right fender. The search eventually brought the bunch of deputies to A. L. Inglesby's garage. No right fenders were bent, but Deputy V. B. Jones, who assisted in the search, accidentally discovered that one car was loaded down with whisky. The back seat of the car had been deliberately sawed out so as to leave room for more booze, and there was enough of the fluid of inebriation to make a company of "Sammies" dream of the surrender of Berlin – enough to make a jackrabbit fight a coyote, or enough to make the entire Bingham police force and the Salt Lake county sheriff's force start on another hunt for Lopes.

Upon inquiry the car was found to be the property of Mr. John E. Creighton. The deputies immediately informed Mr. Creighton of the circumstance. Now, Utah is a dry state and Bingham is a law abiding community and so the officers were obliged to confiscate the contents of Mr. Creighton's car, Mr. Creighton obligingly volunteering to drive the car up to sheriff's headquarters, where the licker could be unloaded.

Mr. Creighton expresses great indignation over the illegitimate appropriation of his car for such purposes and in co-operation with the sheriff's force, is exerting every effort to apprehend the "Huns" who ruthlessly sawed out this back seat "for medical purposes".

The goods taken consisted of some 19 cases of a fine old brew and was worth about $5000. With reference to the quality of the booze, we are no expert on that kind of testimony but can truthfully say that is "smelt" very decent.

The above article was printed in the June 7, 1918 edition of The Press-Bulletin, Bingham Canyon, Utah.

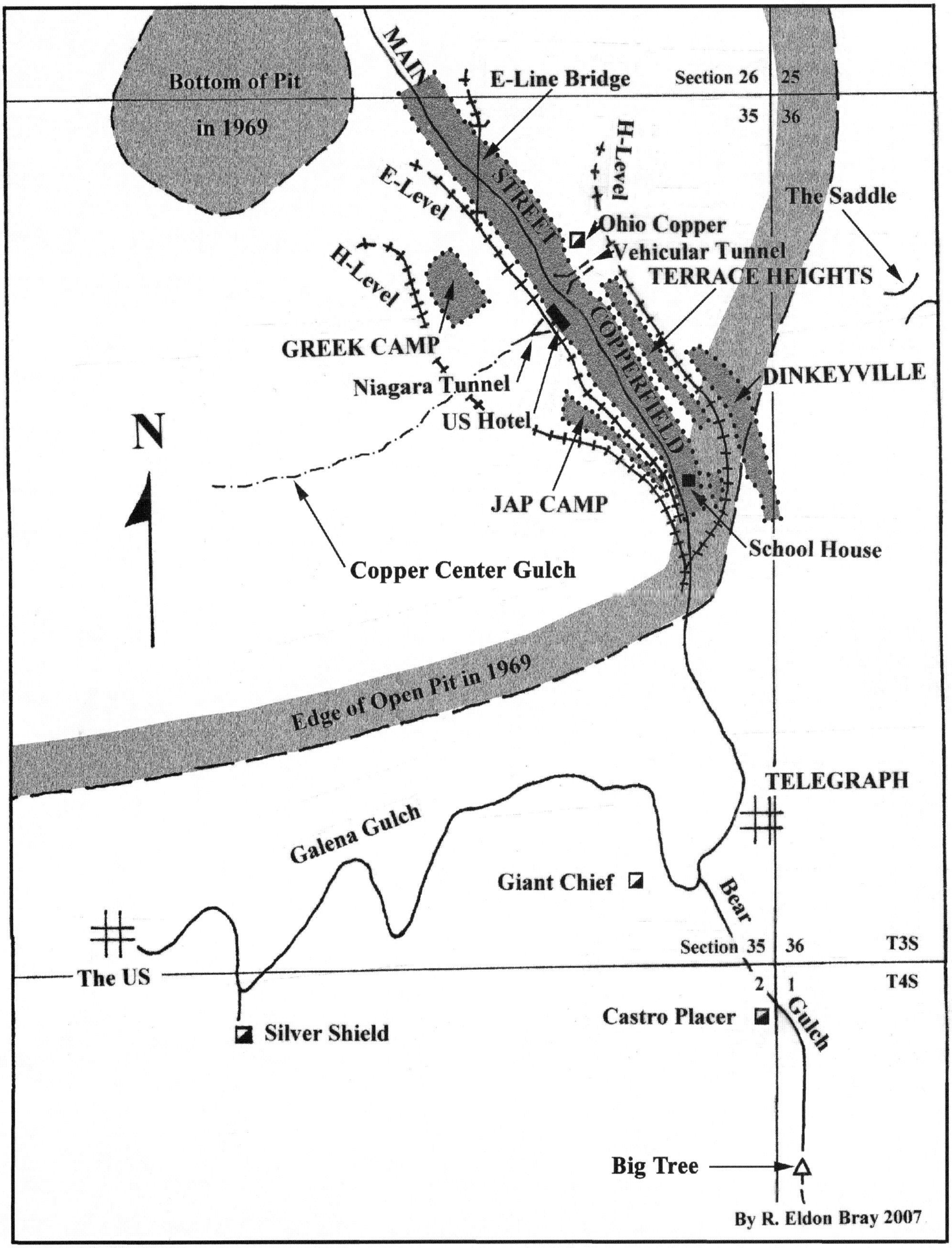

Figure 2 – Location Map of Copperfield Including Telegraph and The US
Drawn by R. Eldon Bray 2007 – The community included the five sections of Copperfield proper

(Main Street, Terrace Heights, Dinkeyville, Jap Camp and Greek Camp) but was also loosely applied to include Telegraph and The US. The roadways and other features are shown as they generally were between the mid 1930s and the mid 1950s. The Bingham-Copperfield Vehicular Tunnel was completed in 1939 and soon afterward the E-Line Bridge and the lower part of Main Street were demolished. Most of the town endured until 1958 when the school was closed and the last buildings were demolished. The edge of the open pit in 1969 indicates the nearly total engulfment and destruction of the town of Copperfield. The pit has continued to expand far beyond the 1969 limits so that, in 2007, a huge hole now occupies the site of the town of which no vestiges remain.

Two of the main levels of the open pit (E and H-Levels) were established immediately after the Utah Copper mine was started in 1906; they provided access from "The Hill" (later "The Pit") to the East Side of the mine.

E-Level and the E-line Bridge went right across the lower part of Copperfield and some people today can remember living virtually below the bridge and feeling their houses shake as trains crossed the bridge.

H-Level was the upper limit of Jap Camp and of Greek Camp and separated the Terrace Heights from Dinkeyville. Residents of the Upper Terrace and of the lower part of Dinkeyville lived with the railroad tracks almost in their back (or front) yards and with the traffic of trains both day and night. The children from Dinkeyville walked across H-Level every day as they went to school or to other sections of the town. As the last days of the town approached the residents contended with the nearby sight and sound of huge electric shovels digging away the mountainside as the mine inexorably closed in on their homes from the north.

Copper Center Gulch wasn't well known to residents of Copperfield because it was filled in with waste rock from the open-cut mine very soon after the mine started operation and so wasn't evident. The mouth of the gulch was marked by an area in the main canyon that was wider than usual. This wide area was occupied by the U.S. Hotel and compressor and by additional houses behind those on the Main Street.

Big Tree was a well-known landmark in the Copperfield area. It was a huge cottonwood tree next to the road in Bear Gulch above Telegraph. It was often a destination for picnics and other outings but we also walked past it on our way to Butterfield Canyon and other places of interest.

The following article, of some interest and amusement, which occurred in Bingham Canyon but perhaps not in Copperfield, was printed in the Salt Lake Tribune on Sep. 1, 1875:

Railroad Blockade

Pat Mahan was arrested yesterday and brought to town on a warrant issued by U.S. Commissioner Sprague for stopping the cars on the Bingham Canyon railroad, which runs through the dump of his mine. Mahan lifted the cars from the track, nailed some planking across it, and filled it in with dirt, making an effectual blockade, and then sat down with his pistol to prevent the removal of the obstructions. He will have a hearing before the Commissioner tomorrow.

From Utah Historical Society Archives

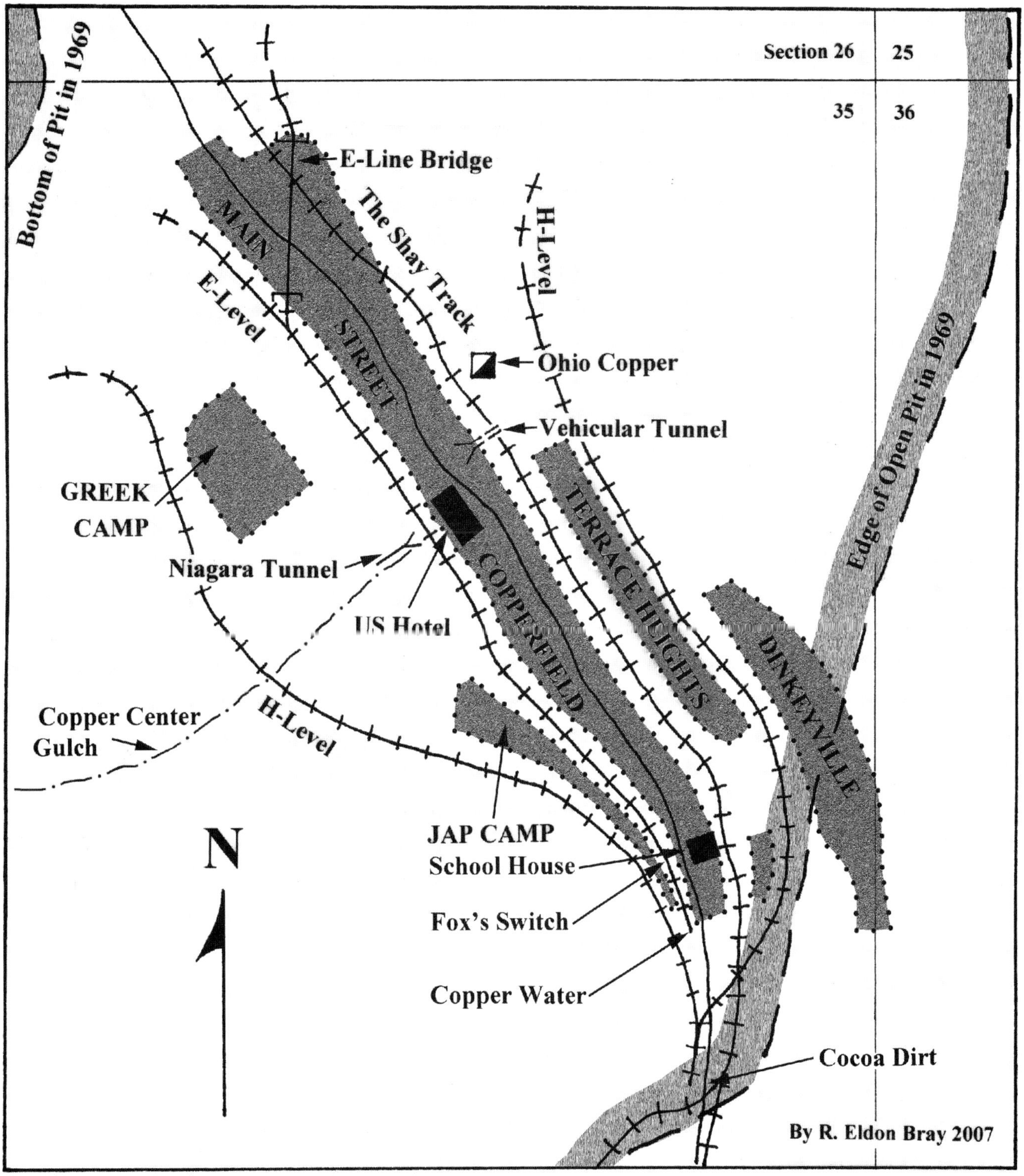

Figure 3 – Map of Five Sections of Copperfield With Roads, Railroads and Other Features Drawn by R. Eldon Bray 2007 – This larger scale map shows more detail of the town and includes the Shay Track which was abandoned by about 1910 and thereafter served primarily as a footpath. The continuation of the Shay Track up into Galena Gulch was, by about 1908, covered by waste rock from the open-pit mine. The down-canyon continuation of the Shay Track, below the E-Line Bridge, was dug away by the open-pit mine as it expanded to the east. E-Level sloped upward from

the E-Line Bridge to Fox's Switch which was between the school house and Jap Camp. Copper Center Gulch is shown on the map although it was filled in by waste rock from the Utah Copper open-cut mine shortly after the mine was started. The Commercial mine was located in the upper reaches of this gulch before it was covered by waste or dug out by the open pit. The wider part of the canyon in the vicinity of the US Hotel was at the mouth of Copper Center Gulch. Fox's Switch, the Copper Water, and the Cocoa Dirt were well-known features of Copperfield.

BINGHAM CANYON BEFORE THE DISCOVERY OF MINERAL

Prior to 1847 the mountain range to the southwest of Salt Lake City was a wilderness known to the Goshute Indians as the Oquirrhs which meant either "Wooded Mountains" or "Shining Mountains". They hunted deer and gathered nuts and berries in these mountains for many generations. Their arrowheads have been found in numerous areas near what we knew as Copperfield – especially by "Big Tree" that was in Bear Gulch and at the old Boy Scout camp in Butterfield Canyon.

The Mormon Pioneers entered the Salt Lake Valley in 1847 and soon began utilizing the resources of the Oquirrh mountains. In 1848 two brothers, Thomas and Sanford Bingham, began grazing cattle and horses in Bingham Canyon and the canyon was named after them.

THE "BIG GROVE" IN SILVER SHIELD

The abundant red pine (Douglas fir) trees in Bingham Canyon were logged by the pioneers and the canyon was especially valued for this resource. The "Big Grove" in Porcupine Gulch (It was known to us as the Silver Shield) was a sturdy stand of especially large trees. Individual trees often measured 3 feet in diameter. Trees from the "Big Grove" were used to build the roof of the Mormon Tabernacle in Salt Lake City. There were early reports of ore minerals being found in Bingham Canyon but the Mormon leader, Brigham Young, discouraged prospecting because his people's greatest needs at that time were for more basic items such as food, shelter, and clothing. For this reason the "Big Grove" was more important to the earlier settlers than the showings of lead-silver mineral in this same gulch and in others nearby.

Gene Halverson's memories of the "Big Grove" in the early 1940s are of coming across huge stumps that were 3 or 4 feet across. He never did measure their diameter. They were cut close to the ground on the upper side so they didn't stand very high. Some were so big that he even laid down on one (not stretched out of course). He looked at those huge stumps and wondered at their size and puzzled over how the lumbermen got them off the mountain or in how many pieces. There was a lot of thick brush in the area and there were several caved-in mine tunnels and small shafts that were almost hidden by the brush. A person could easily fall into one of these openings so you had to be very careful of where you walked. Gene lost his dog while in that area – he thought it could have fallen down a mine shaft.

Does anybody have a photo of the Copper Belt Railroad Bridge crossing Upper Copperfield or of the railroad in Galena Gulch (about 1904)?

Figure 4 - Silver Shield (Porcupine) Gulch – Date Uncertain

From the Gordon Bodily Collection - Silver Shield (Porcupine) Gulch was the site of the "Big Grove" of red pine (Douglas fir) trees used to build the roof of the Mormon Tabernacle in Salt Lake City. The upper, right-of-center area where evergreens can be seen is where stumps of huge trees remained more than 80 years later. A dozen or so old mine dumps, including the biggest one of the Silver Shield mine can be seen in this photo. There were numerous other smaller mine workings hidden by the brush. This area, peripheral to the open-pit copper mine, contained minerals of lead, zinc, silver, gold, and other elements including arsenic. The mines exposed these minerals to the air so that they oxidized and some of these elements were leached into the ground and surface water. (See "Shower Where You Can"). In the 1940s we hiked from Copperfield to the top of this gulch to cut Christmas trees. The bottom of the main (Galena) gulch has been filled in with waste rock from the Utah Copper open-pit mine which covered the railroad and the roadway that originally was near the bottom of Galena Gulch. A new roadway was constructed higher up the hillside; it can be seen in this photo; it goes just below the Silver Shield mine.

THE DISCOVERY OF MINERAL IN BINGHAM CANYON

In the fall of 1863 George B. Ogilvie found lead-silver mineralization in the upper part of Bingham Canyon known as Galena Gulch (the old U.S. above Copperfield). On September 17, 1863 the West Jordan Claim was formally located by Ogilvie and twenty-four others. This was the first recorded mining location in Utah. Another claim was located the next day and other claims soon followed.

The Jordan Tunnel was started in 1864 by the West Jordan Mining Company. The outcroppings of high-grade galena (or, if oxidized, of lead carbonate) were very rich in lead and silver but the costs of transportation and supplies were so extremely high that the mines couldn't be operated at a profit. The coming of the railroad was anticipated within a few years and the costs of transportation and supplies were expected to drop at that time. It was thought that many of the claims would then be profitable. The mining laws were revised so that the claims would remain valid until after the coming of the railroad.

The Union and Central Pacific railroads were completed in 1869, the Utah Central connection with Salt Lake City was made in 1870 and the railroad to Bingham was completed in 1873. In response to the improvement in transportation and the reduction in other costs, underground mining activity was markedly renewed in 1870 and continued to accelerate for several years. Known production of silver in the Bingham District, from 1870 to 1900, peaked in 1885 with 1,350,000 ounces.

LETTER TO LADY LOVE

A Bingham fellow received a letter a few days ago from his lady love who left town a few weeks ago to take up residence in a distant state. The letter was such a nice one and he was so anxious to answer it in like mind that he took into his confidence two Bingham girls, who are well versed in the gentle art, and requested them each to write a reply, promising that the one turning in the best letter would be rewarded with a box of candy. The girls did, and the letters they wrote were dandies, and the girl at ________ will have some interesting reading matter about Sunday. Copies of these letters will be published next week.

The above article was printed in the March 22, 1918 edition of The Press-Bulletin, Bingham Canyon, Utah.

THE OLD JORDAN and GALENA MINES

Figure 5 – The Old Jordan Mine in 1900
From U.S. Geological Survey Professional Paper 38 - The Old Jordan Mine, site of the first recorded mining location in Utah. The original discovery was made at right of middle foreground. This view shows surface workings which penetrate the Jordan limestone in fore- and middle ground and the old upper adits of the Commercial Mine on footwall of Commercial limestone at right in background. The Old Jordan Mine was started 36 years before this photo was taken. All or parts of the buildings and other structures were later demolished so that individuals who lived at the U.S. in 1931 were unable to identify most of the buildings in this photo. Perhaps old mine dumps in 1931 could have been matched up to The Old Jordan Mine.

The "U.S." as we called it when I was growing up included essentially all of Galena Gulch but especially the sites of the Old Jordan and Galena mines. We didn't hear of or know the proper name of Galena Gulch but instead referred to this area by the name of the mine or the name of the company that owned the mine that was active there. I can barely remember there being some buildings up at the U.S.. By 1945 all the buildings were gone and only the waste dumps remained. Deer and other wildlife were plentiful but no people lived there.

Does anybody have a photo of the metal "Gun Turret" that was on the hillside above the U.S. Mine in Galena Gulch? It was built for protection against Strikers.

Figure 6 – The Old Jordan and Galena Mines About 1909
From "Bingham, The Greatest Copper Camp In The World" 1909 – This photo shows several buildings that were still in place in 1931 at the Old U.S.. Bingham Canyon view looking west of the Old Jordan and Galena mines of the United States Smelting, Mining & Refining Co. These mines have poured out vast treasures in gold, silver and lead during the last twenty-five years and are active producers today (1909). The ore is taken by aerial tram to Bingham station, where it is loaded into railroad cars for the company's smelter. Four of the "stations" for the aerial tramway are visible in this photo. The aerial tramway came into use about 1902 and was discontinued in 1910 after the D&RGW High Line extended rail service to the U.S. mines. The rail service was discontinued within a few years because the open-cut mine either dug out the tracks or covered them over with waste rock.

PLACER GOLD MINING

In early 1864 a party of old Californians found free gold in the gravels of Bingham Canyon. Virtually everybody recognized placer gold, just as it came from the gravels, as a precious commodity. It was of immediate value to its finders since it could be sold or traded for goods without the need for costly transportation or processing. Gravel washing for gold was actively taken up in the spring of 1865, the year following its discovery.

During the lull in underground mining from 1864 to 1870 there was consistent activity in placer mining for gold. The years of maximum placer mining were from 1868 to 1872. Up to 1871 about

$1,000,000 in gold was recovered from the gravels of Bingham Canyon. The price of gold in 1900 was $20 per ounce. At a price of $500 per ounce, which is close to today's prices, that amount of gold would be worth over $25,000,000! One nugget that was found near Damphool Gulch (2,000 feet above Dry Fork) in 1876 weighed 7 ounces and 15 pennyweights; it was thought to be the largest gold nugget ever found in the state of Utah.

Figure 7 – A Placer Gold Mining Operation Below Copperfield About 1904
From U.S. Geological Survey Professional Paper 38 – View is north-northwest (downstream) of auriferous creek gravels in Bingham Canyon below Upper Bingham. The creek gravels, with considerable plant roots, have been dug down presumably to bedrock where the gold is generally most abundant. Three sluice boxes can be seen in this photo – two in the center foreground and a smaller one farther back near the bend in the stream.

Placer gold was mined from the old creek and bench gravels for most of the length of Bingham Canyon with very good success in some places and not so good in others. Most of the operations used sluice boxes into which the gravel was shoveled by hand. Shafts were often sunk down to bedrock and then tunnels were run along the top of bedrock. One such tunnel that extended up both Carr Fork and Bingham Canyon was said to be 1,500 feet in length with a shaft every 250 feet that went through the 60 feet of overlying debris. Some bench gravels that were 375 feet above the creek between Carr Fork and Dixon Gulch were hydraulically mined. An 80-foot stream of water could be thrown from a 3-inch nozzle. The deposit from surface to bedrock for a distance of 200 feet was cut out by the water and led into a sluice in Dixon Gulch where the gold was recovered.

THE CASTRO PLACER IN BEAR GULCH

Figure 8 – Castro Placer in Bear Gulch About 1900
From U.S. Geological Survey Professional Paper 38 - The open cut was down to bedrock and tunnels were driven from the open cut to pay dirt at the base of the gravels. There is a sluice box left of center in the bottom of the creek bed. Water was scarce so dams or water holes were used to trap the water in the springtime. The deposit was then operated several times over the years for a few months while water was available. This operation was about 300 yards below "Big Tree" where a small man-made pond was located in the 1940s. This was also the site where my friend Warren and I recovered some placer gold in the 1940s (see "Gold At Bear Gulch").

Near Copperfield, when I was growing up, there were a couple places with historical evidence of the old placer mines. The pond of water in Bear Gulch a couple hundred yards below Big Tree (just about the area of this photo) was the site of a placer deposit that was worked from 1868 to 1872 by a miner named Castro. It was named "The Castro Placer" but I don't recall hearing that name while I was growing up. Because there was no year-round stream in Bear Gulch the water was impounded at this location expressly for the purpose of placer mining each springtime. In 1873 it is recorded that $27,000 of placer gold were mined from Bear Gulch. At $500 per ounce this would be about $653,000. Intermittent placer mining continued in this area into the mid 1940's. Alvin Cole was one of the last known individuals to mine placer gold from this location.

There was a vertical shaft in the middle of "Jackass Gulch" (the gulch the mouth of which was a few yards below Big Tree). The shaft was filled with sediment and water and had some old tree limbs laid across it so that people wouldn't fall into it. I wondered why it had been dug there. It is now obvious that it was dug in hopes of finding placer gold. Placer gold is very heavy and, after being eroded from the country rock, is often concentrated in the deepest part of a stream bed or even in cracks that extend down into the bedrock. If good amounts of gold are found in such a shaft then cross cuts or tunnels are often dug from the bottom of the shaft out along the gravels just above bedrock. If there were crosscuts going out from the bottom of this shaft then the dump should have been extra large but I don't recall whether that was the case. There was another such shaft by the creek in Butterfield Canyon.

These shafts and tunnels in unconsolidated gravels were prone to cave-ins and so were very dangerous. It is reported that the three Freeman brothers were killed in a cave-in of placer diggings at the mouth of Freeman Gulch.

MORE UNDERGROUND MINES

After the Utah Central Railroad was connected to Salt Lake City in 1870 the costs of machinery and supplies to the underground mines in Bingham were much improved and several new underground mining ventures were undertaken. However, there was still a need to further reduce the transportation and haulage costs between Bingham and Salt Lake. The completion of the railroad to Bingham in 1873 eliminated this last obstacle for underground mining in the canyon. Transportation of ore and materials was assured, the costs of machinery and supplies were reduced still more and the Latter-day Saints (Mormons) were now encouraged by their authorities to participate in mining enterprises. The results were the location of several more mines and the exploitation of others that had been awaiting more favorable economics.

There were several epochs of mining different types of ore. They were (1) the epoch of lead carbonate, (2) the epoch of oxidized gold ore, (3) the epoch of sulphide lead and (4) the epoch of copper mining.

The epoch of lead carbonate mining was first – the early, near-surface ore bodies were mainly of lead carbonate with silver. However, several mines as they proceeded to greater depth in 1874, encountered lead sulphide ore that required a different milling method that had not yet been developed. Upon encountering the lead sulphide ore several mines turned their attention to oxidized gold ore.

The epoch of oxidized gold ore began in the late 1870s and early 1880s following the temporary exhaustion of lead carbonate ore bodies. By the close of 1882 four stamp mills had been erected to process oxidized gold ores. This period of mining was of questionable profitability and ceased within about ten years.

The epoch of lead carbonate mining was then extended by the development in the early 1880s of large lead carbonate orebodies in Butterfield Canyon and at the Yosemite and Brooklyn mines.

An epoch of successful silver-bearing lead sulphide mining commenced in about 1884. In 1891 and 1892 there were 21 producing mines of which the Old Jordan and Galena, Brooklyn, Telegraph and Yosemite (in the Copperfield area) were among the top seven producers. In 1893 governmental actions caused the price of silver to drop and mine production was drastically curtailed.

The epoch of copper mining essentially began in 1896. Sulphide copper had been encountered many years earlier but it had not been considered to be valuable. However, uses of copper were growing and the price of copper increased. In December, 1896 5,000 tons of copper ore were shipped from the

Highland Boy property and copper ore became regarded as a valuable commodity. The period of copper sulphide ore was then begun and it soon proved to be the greatest of all the ore types to be exploited.

It was also becoming apparent by 1896 that the consolidation of several small adjoining properties could result in lower mining costs. There were a number of mines which had encountered low-grade copper that was of doubtful profitability; they could especially benefit from consolidating by which they would lower both their mining and their smelting costs.

In 1899 the United States Mining Company was formed and consolidated the Old Jordan and Galena, Niagara, Old Telegraph and other mines in the Copperfield area. The U.S. was the only major underground mining company that was active in Copperfield while I was growing up. Some leasers continued to mine a small amount of ore in the U.S. and Telegraph areas and to haul the ore by truck down to the ore cars at the upper end of Copperfield by Fox's Switch.

SILVER HILL AND PORCUPINE (SILVER SHIELD) GULCHES

Figure 9 – Silver Hill (left) and Porcupine (Silver Shield) (right) Gulches About 1909
From "Bingham, The Greatest Copper Camp In The World" 1909 These were side gulches of Galena Gulch about halfway between the Old U.S. (Old Jordan and Galena) and Telegraph. They had many old mine dumps as did many of the gulches in and around Bingham Canyon. The Silver Shield was the most recently active mine in these two gulches and the main road to the U.S. went just below the Silver Shield mine so we were most familiar with it. The Silver Hill Gulch was seldom visited and was largely unknown to us. Several trails or roadways can be seen in this photo; they were largely grown over by brush in the 1940s.

The Silver Hill claim was located in July of 1864. The ore was mostly lead-silver ore with some gold. Carbonate lead graded downward into silver-bearing lead sulphides. The Silver Hill vein was mined through three tunnels called "Three Black Dog Tunnels". The old mine dumps that were in a line, one above the other, were generally at different elevations along one vein such as the Silver Hill. The Silver Shield was a famous producer and a mill was built there. The vein at the Silver Shield was also tapped at several elevations that resulted in several old dumps all in a line.

THE OLD TELEGRAPH MINE

Figure 10 – The Old Telegraph Mine About 1909
From "Bingham, The Greatest Copper Camp In The World" 1909 – The Old Telegraph Mine building and bunkhouse. The Old Telegraph Mine was a group of mines that included the Evans, Tribune, Grecian Bend, Roman Empire and others. The Old Telegraph, Spanish, Galena, and Old Jordan mines lay along a famous limestone ore zone known as the Jordan Line. The four mines were secured in 1899 by the United States Mining Company. After many old surface mines were abandoned, active mining of the same orebodies was often continued from tunnels at lower elevations. The Niagara, Butterfield, and Mascotte tunnels allowed water to be drained down to the tunnel level rather than costly pumping to the surface and mining was simplified by pulling the ore down to the tunnel levels without the need to hoist the ore up through shafts.

This was the site of the Telegraph section of Copperfield as we knew it when growing up. Practically all of these buildings had been demolished and newer homes constructed on the upper dump by the time I was familiar with Telegraph.

All the structures of the lower mine were gone before my time. A stamp mill that crushed and ground ore to the consistency of fine-grained sand was built and operated on the mine or dump site. The mill might have been built either before or after the mine structures were removed. All the structures of the mill were also gone before my time. The dump was leveled off and was covered with the soft tailings sand from the mill so that it was an ideal ball field. The kids from Dinkeyville and Copperfield used to come up here to play football and baseball. The nearby "cocoa dirt" was the fine-grained reddish brown tailings from this same mill that apparently treated oxidized, brown gold ores.

The Old Telegraph mine included a large number of claims on both sides of Bear Gulch. A body of ore that was encountered in 1874 on the Nez Perces Chief and No-You-Don't mines was also claimed by the adjoining Montreal. This orebody became the subject of litigation. As a result of the litigation and purchases a group of these claims was consolidated under the name "Old Telegraph" mine. In 1878 suit was brought against the mine superintendent for alleged misrepresentation of facts to depreciate the market value of the stock. The superintendent ended up paying the company $200,000. The property came into the control of the U.S. Company in 1899; at this time the total output of the mine was estimated to have been about $16,000,000. The ore in this group of mines varied from gold to silver-lead with some copper-iron. Some assays showed as much as 230 ounces of silver per ton. The famous Black Stope ore was reported at 45% lead, 18% silver and 1.5% copper.

The BINGHAM MINES CO.

The Bingham Mines company has just completed another successful year under careful and capable management. They have been able to operate uninterruptedly for the past year, regardless of the low prices of the various metals.

The Bingham Mines property includes the Dalton & Lark mine, Lead mine, and Brooklyn mine in the eastern part of the West Mountain Mining district, and the Commercial mine in the center of the same mining district. These mines have been steady producers of lead-silver and copper ores for the past forty years, which demonstrates the remarkable persistence, size and value of the ore shoots. They have attained a depth of 2,000 feet on the Brooklyn ore shoot and a depth of 1,350 feet on the Lead mine and Dalton & Lark ore shoots, respectively. These mines are developed by drifts from the Mascotte tunnel, which is operated by the Bingham Central Railway company. They have attained a depth of 1,300 feet on the Commercial ore shoot, which is being developed through the Niagara tunnel. The product from these mines is shipped direct to the smelters.

Mining and development work has been conducted in a most efficient manner, which is reflected in the present splendid physical condition of the different mines.

Mr. James P. Graves of Boston is president of the company, Mr. Imer Pett, general manager, with offices at 404 Dooly block, Salt Lake City, and Mr. Joseph Hyland, general superintendent, with headquarters at Lark, Utah.

The above article was printed in the January 1, 1915 edition of The Press-Bulletin, Bingham Canyon, Utah.

THE BROOKLYN MINE IN YOSEMITE GULCH

Figure 11 – Outcrop of Quartzite Foot Wall of Brooklyn Ore Bodies About 1904
From U.S. Geological Survey Professional Paper 38 - The Brooklyn mine was in Yosemite Gulch 600 feet southeast of the Yosemite mine which was to the left of this photo. In the 1940s the Yosemite mine and its dump were more prominent than those of the Brooklyn. The trail just below the rock outcrops led to the saddle above Copperfield. We kids from Copperfield used to hike to the Yosemite mine quite often. This general area is where many of our adventures with deer and porcupines took place. "Bishop", the watchman at the Yosemite mine, trapped a bobcat beneath one of the rock outcrops shown here. The electrical substation with the barbed wire fence where the wild housecat bit me was just to the left of this photo. The tunnel (see "Dangerous Shortcut") that we kids went through was also to the left of this photo.

When was the Upper Crossing (Wooden H-Level Bridge) demolished and replaced by the concrete overpass? It was still there in 1916.

THE OHIO COPPER SHAFT

Figure 12 – The Ohio Copper Shaft in Upper Bingham About 1909
From "Bingham, The Greatest Copper Camp In The World" 1909 – The headframe of the Ohio Copper shaft is directly above the locomotive on the bridge. Ore from the Ohio Copper was mined by underground methods and hauled out to Lark through the Mascotte tunnel. Very little of its mining activities were evident in Copperfield. The East Line Extension Bridge, on the right of the photo, crosses the canyon and connects the main, or West Side of the Utah Copper mine to the East Side on E-Level along which waste rock was hauled to the dumps. The highest level shown, above the Ohio Copper, is "H-Level". Another bridge, the Upper Crossing, or wooden H-Level bridge is farther up the canyon to the right of this photo; it also provided access from the West Side to the East Side of the mine. Dinkeyville and Terrace Heights weren't built yet.

The Ohio Copper shaft wasn't well-known to us as we grew up in Copperfield because it was somewhat out of the way. The Ohio Copper Company came into existence in 1903 when it purchased the Columbia mine which had shipped 175,000 pounds of copper in 1902. The Ohio Copper Company went on to become one of the early large-scale copper producers in Bingham Canyon. The ore was mined underground but wasn't hauled to the surface in Copperfield. The Ohio Copper built a 4,500 ton-per-day mill on the other side of the mountain by Lark and hauled the ore out to Lark through the Mascotte tunnel. Signs of the mining activity and mill operation were most obvious over in Lark where the concrete foundations of the mill were still visible in the 1950s and the mill tailings were known to us as "the sand dunes".

The Ohio Copper shaft was below H-Level and above Stella Klopenstein's house and the Lendaris Merc. The water tank (Stella's drinking water) was near this shaft and the pipeline where the kid shredded his pants was also near the shaft.

THE HOLDEN TRAMWAY

Figure 13 – Ore Cars Coasting Down Upper Canyon, About 1892
Used by Permission, Utah State Historical Society, all rights reserved - These ore cars were rolling down the Holden Tramway by gravity. Each of the men is holding a rope that is attached to a brake lever. Empty cars were pulled back uphill to the underground mines by mules.

The Holden Tramway was built in the early 1890's by L. E. Holden. It was a narrow-gauge (20-inch) mule and gravity tramway that was used to haul ore from the Jordan, Galena, and Old Telegraph mine areas. It was 3 miles from the Telegraph mine down to the Rio Grande Western standard gauge rail station in Lower Bingham where the ore was loaded into Rio Grande trains and hauled to the smelters. The gradient of the mule tramway was from 4 to 14 percent which is very steep and upon which the trains could, if unchecked, attain a considerable speed. The men riding these cars are each holding a rope by which they apply the brakes to slow or stop the train. It would be interesting to know if or how many runaways they had as the ore trains came down the canyon and what the results were. One muleskinner said 24 cars derailed in one day. After the ore was unloaded the empty cars were pulled back up to the mine by mules.

Figure 14 – Mules Pulling an Ore Car Up the Holden Tramway About 1892
Used by Permission, Utah State Historical Society, all rights reserved -These mules were the most practical if not only way of getting the empty cars back up the hill.

The mules were then returned, in groups of ten or twelve, to the lower end of the tramway where they were stabled.

When we were growing up the "Cocoa Dirt" was an embankment of reddish brown clay-like material between the road and the tailtrack of H-Level just below the "ball field" in Telegraph. The Cocoa Dirt was used by the kids in Copperfield much like modeling clay is used today. Across from and above the road from the Cocoa Dirt was a trail that went from Telegraph to Dinkeyville. This trail was the old roadbed of the Holden Tramway. The rails and crossties had been removed so there was no evidence of its prior utility. The trail had an even grade and was easy walking but there was a steep dropoff down to the roadway so you had to be careful not to get close to the edge. Some kids referred to it as "The Telegraph Trail" while to others it just didn't have a name.

Does anybody know the details of the Trackshifter Operator who insisted on going up the switchback and had to jump off before getting hit by a train of loads?

Figure 15 – Copper Belt Railroad train at the Telegraph Mine about 1902
Used by Permission, Utah State Historical Society, all rights reserved – 1901 was the first year that Shay engines were used on what had previously been the gravity and mule-driven Holden Tramway. The Holden Tramway became the Copper Belt Railroad in about 1902. This engine appears to be very new.

In 1899 the Holden Tramway was purchased by J. G. Jacobs. He extended the line and converted it over to the use of Shay engines which started hauling ore in February of 1901. The Shay engines were capable of pulling loads up steep grades and, being smaller than other steam engines, they were able to operate on narrow-gauge tracks. However, the Shay engines were slower and couldn't haul as much as the larger engines by which they were later replaced. After conversion to Shay engines the Tramway was purchased by William Bagley who reorganized the company as The Copper Belt Railroad and extended it up Carr Fork.

The "Shady Track" (as we called it) that ran along the hillside above the vehicular tunnel, the houses of Main Street and the schoolhouse in Copperfield was the abandoned roadbed of the Shay Track of the Copper Belt Railroad. It wasn't especially shady and at the time I couldn't understand why it was called the "Shady Track". Photos of Terrace Heights and of Main Street from the Niagara tunnel area show the old Shay Track especially well.

The Holden Tramway, the mules and the Shay engines on the Copper Belt Railroad, photos of which are shown above, all preceded the development of the open-cut copper mines of the Boston Consolidated and the Utah Copper companies. Shay engines that were larger and of a wider gage continued to be used for a few years after the start of the open-cut mines. These Shay engines were soon replaced by still larger steam engines that were capable of higher speeds and of pulling heavier loads.

Figure 16 – Shay Engines Going Up a Steep Grade About 1904
Used by Permission, Utah State Historical Society, all rights reserved – These engines belonged to the Copper Belt Railroad. The Copper Belt was a narrow-gauge railroad with a grade of 6 to 7 percent; Shay engines were well-suited for these steep, narrow grades but were very slow. The Copper Belt hauled ore from the underground mines from 1902 to 1907 and from the open-cut mines from 1906 to 1907. This photo is not of a level in the open-cut and the grade is very steep. A "guesstimate" is that this photo was taken about 1904. The D&RGW High Line replaced the Copper Belt in 1907.

During the first 50 years of mining in Bingham Canyon there were half a dozen or so different railroads and tramways. Several of the tramways were aerial and some of them interconnected with railroads or even mule-powered tramways.

Some early photos show the Holden Tramway with a snowshed-like covering over the railway while in other photos there is no covering. Could this presence or absence of the covering be related to the use of steam engines whose size and operating requirements precluded the use of the covering?

THE "COPPER HILL" BEFORE THE OPEN-PIT MINES

Bingham Canyon is about seven miles long from its head in Galena Gulch (the old U.S.) to its mouth near Copperton. The "Copper Hill" as it came to be known, was the mountain on the southwest side of the canyon that extended almost two miles from Carr Fork to Galena Gulch. It rose quite steeply to a ridge about sixteen hundred feet above the creek bottom.

Figure 17 - Upper Bingham and The Copper Hill in 1904
Photo Courtesy Of Kennecott Utah Copper - The Copper Hill takes up the right half of the photo. Almost 40 years of vigorous prospecting and underground mining had already taken place. Dozens of mine dumps and several well-defined trails, roads, or railroads can be seen on the hill. Numerous buildings and several large underground mines and processing mills are evident along and near the canyon bottom. Large-scale open-cut mining had not yet been started. Carr Fork is just off the right edge of the photo. Copper Center Gulch enters the main canyon from the right three-fourths of the way up the canyon bottom in this photo; it cuts back into The Copper Hill for half a mile. The Copper Center Tunnel, with a cluster of structures, is near the mouth of Copper Center Gulch. Galena Gulch, the main fork of Bingham Canyon, or the U.S. as we called it is the largest gulch that comes in from the right farthest up the canyon. Copperfield, as we knew it, occupied the upper one-fourth of the main canyon bottom from 500 feet below the mouth of Copper Center Gulch to the mouth of Galena Gulch with Telegraph (and the U.S.) being farther up. Essentially none of the structures that are visible in this photo were still there in 1945. The Utah Copper open-pit copper mine, when I was growing up, had mined away everything in this photo below Copperfield. "The Copper Hill" had become "The Pit".

Prior to the 1850s or 1860s there had been stands of large red pine (Douglas fir) trees on the "Copper Hill" and surrounding mountains. With the coming of the Mormon Pioneers in 1847 the Oquirrh Mountains were turned to for a source of timber for the construction of homes and other buildings. With the discovery of mineral in Bingham Canyon in 1863 the mining industry was started and there was a need for mine timbers as well as for more buildings in the canyon. The result was the cutting of all the large trees in Bingham Canyon and in the surrounding canyons for miles around.

By 1904 the "Copper Hill" had been subjected to almost forty years of timber cutting, prospecting, and mining. Now the mountain was covered with brush and only small trees. Half a dozen trails or roads for dragging timber or hauling ore had been cut through the brush and rocks. The mountain was also scarred by the shafts, adits, dumps and ore bins of a couple dozen small underground mines.

Most of the larger underground mines were located around the periphery of the Copper Hill and produced high-grade lead-silver or copper-gold ore from veins or limestone replacement deposits. Low grade copper ore had been encountered on the Copper Hill but was considered in those early years to be unprofitable. Now, however, interest was growing for large-scale exploitation of this resource -- properties were being purchased and consolidated. Several underground mines were driven into the hill and sampled to determine the potential of the huge lower-grade body of copper ore.

The topographic map and the generalized sketch map (below) of the Bingham Canyon area before the advent of open-cut copper mining show the major canyons and gulches with the railroads, tramways, and larger underground mines located on them.

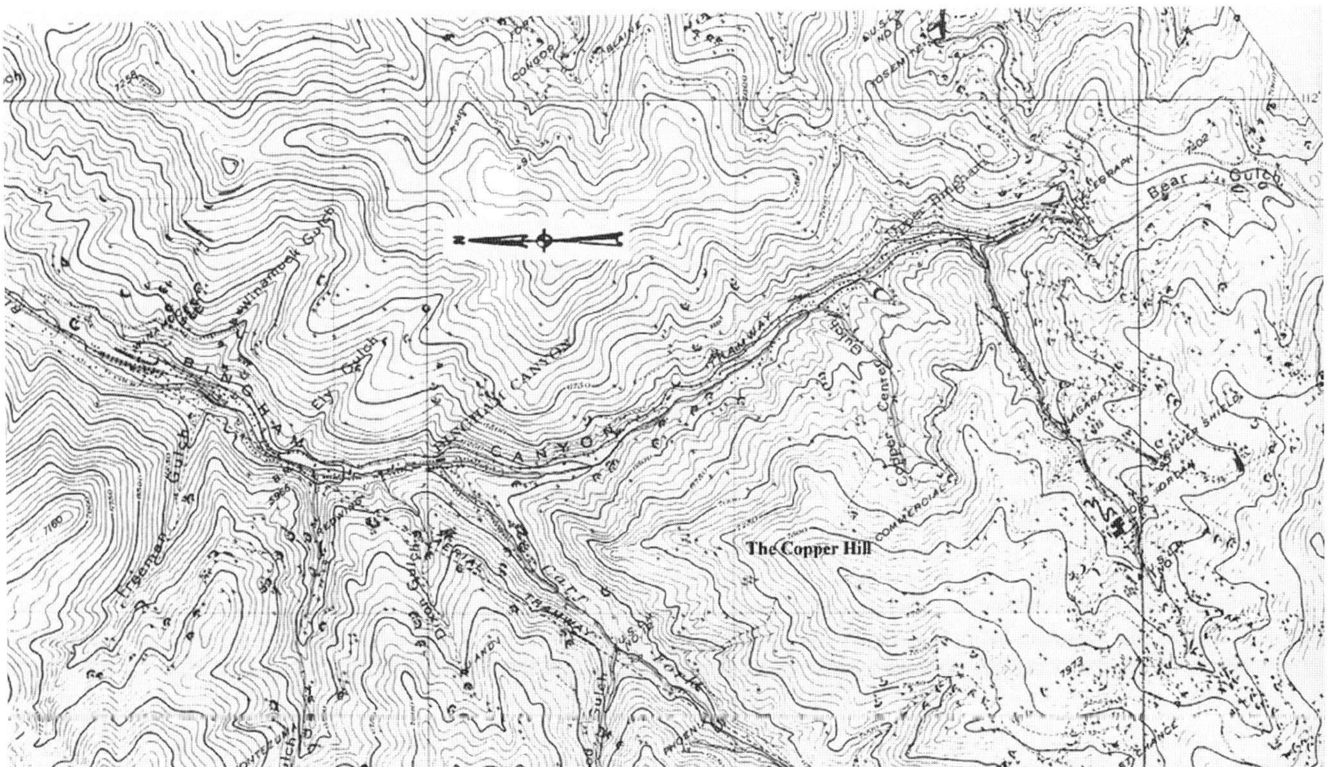

Figure 18 – Topographic Map of Bingham-Copperfield Area About 1900
From "Utah Ghost Rails" by Carr & Edwards (with modification of North arrow) – Almost the entire area of this map is now occupied by the Utah Copper open-pit mine or is covered by its waste dumps. The mouth of Carr Fork is immediately below the North arrow and runs to the southwest off the lower part of the map. Galena Gulch (the U.S. as we knew it) is the large canyon on the

right side of the map; it also runs southwest (to the lower right corner of the map). The "Copper Hill" is the mountain between Carr Fork and Galena Gulch. "Upper Bingham" as noted on the map became known as Copperfield.

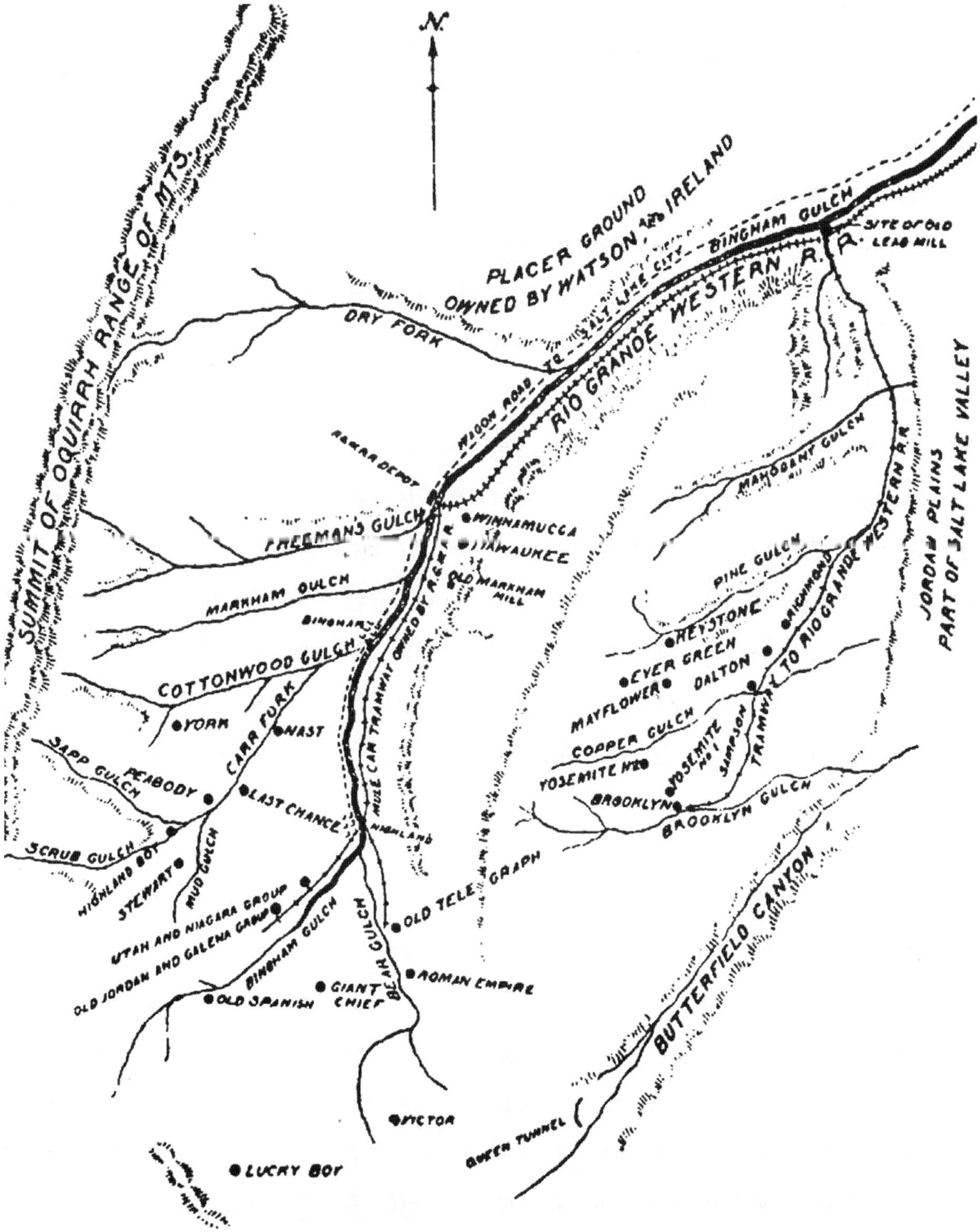

Figure 19 – Map of Bingham Canyon Showing Major Mines and Railroads About 1899
From 1899 Issue of "Mines and Minerals", "Old Reliable" and the Eugene Halverson Collection – There are no indications of the "Copper Hill" and the upcoming open-cut (later referred to as open-

pit) copper mines. Many of the names shown here were unfamiliar to us. We knew Bingham Gulch (Galena Gulch) as "The U.S.", Pine Gulch as "The Fortuna or Keystone Gulch " and Brooklyn Gulch as Yosemite Gulch. Mahogany Gulch had no name and was seldom visited by anybody from the area. The Mule Car Tramway was upgraded, outfitted with Shay engines and became known as the Copper Belt Railroad and "The Shay Track". The "Tramway to Rio Grande Western RR" that traversed the foothills from Lead Mine to the Brooklyn mine consisted in the 1940s of unnamed roadways or trails that had been largely obliterated by the construction of the High Line of the Rio Grande Western Railroad (which we knew as the Horseshoe Bend).

A REAL GHOST WAS SEEN IN BINGHAM

According to talk, among the employees there is now and has been for two weeks a ghost in the camp. A real, strange, mysterious ghost that has caused many who have seen and heard of it to wonder at its queer capers. Heretofore this community has not been bothered much with ghosts and "haunts" and there are doubtless many people here who have never seen one and it seems strange that a common ghost would pick out a mining camp as a field for operation. And there are no negroes here with wild, fanciful imaginations to take fright at the appearance of the supernatural. This lack of other places of abode, but in spite of all this a ghost has paid this camp a call.

The alleged ghost made his debut some two weeks ago on the Utah Copper hill and his first appearance in public created quite a little excitement among those to whom he favored with a call. According to those who have seen it or heard what the other fellows said who did see it, the ghost is a handsome, well dressed man with a rather long nose and somewhat chiseled features. He was first observed one night coming from O level down to N. He had no lantern and he did not seek out the trails, but came straight over the precipices and reaching the level approached a shanty where a few souls were assembled. The men did not know it was a ghost and started towards it, thinking it was a man, as it looked. At the sight of the men the ghost began a flanking movement, ran across the level and jumped over a high and dangerous precipice to the M level. Those who saw it were horrified. They decided it was a crazy man or some desperate criminal and they knew he was killed. So they picked their way around by a trail to the lower level to find the mangled remains of the unfortunate, but they found nothing. They were puzzled at the appearance and the disappearance of this strange visitor, but had this been the only call it might have passed unnoticed. It wasn't. Mr. Ghost came again and seemed to take special delight in letting the men chase him awhile and then he would jump over a steep cliff and disappear. These capers he kept up night after night, but would never enter into conversation with the men, and would disappear when it seemed that he was about to be cornered. And when he disappeared there would be no trace of him left behind.

But at last one night he descended to the N level and the men there decided to capture him alive. When the men started for him the ghost fled as usual, but this time he did not choose to jump from one level to the other and ran along the level with the men in hot pursuit. The men gained on him and just as he was approaching a sharp curve on the level were almost ready to lay hands on him, but when the ghost rounded that curve he vanished into thin air. It is said that there was no possible way for him to escape or hide there, but when the men reached the corner they could not see hide nor hair of him.

There is a diversity of opinion about this apparition, but in view of the fact that the ghost is still at large we can see no necessity for elucidating upon mere opinion.

The above article was printed in the January 11, 1918 edition of The Press-Bulletin, Bingham Canyon, Utah. Used by Permission, Utah State Historical Society, all rights reserved.

CHAPTER TWO

THE UTAH COPPER COMPANY

The Utah Copper Company was not the first to begin mining low-grade copper ore from "The Copper Hill". Nor was it the first to begin mining on The Hill with open-cut methods. Both those distinctions belong to the Boston Consolidated Mining Company.

Figure 20 - Utah Copper Mine in August 1906 When First Steam Shovel Was Started
From "Bingham, The Greatest Copper Camp In The World" 1909 – This view of the beginning of the open-cut Utah Copper mine is looking up the canyon. An underground mine, with a wooden cover over the outside track is also shown – it was used to sample the ore and to provide ore for milling tests before the open-cut operation. The open-cut mine was started well above the creek bottom and pushed its way back into the mountain.

The Boston Consolidated Mining Company purchased and consolidated a large number of claims in 1897 that included operating properties on which high grade copper ore was being mined and also claims that contained only low grade copper ores. Many of these claims were on the area that would later become part of the Utah Copper open-cut mine. The Boston Consolidated continued to mine the higher grade copper ore from nearby underground mines but a few years later also started mining lower grade copper ore from the upper part of the "Copper Hill", still by underground methods.

In mid 1906 the Boston Consolidated Mining Company began mining lower grade copper ore by the open-cut method on the upper reaches of the "Copper Hill" west of the main road of Bingham Canyon with four steam shovels. The ore and waste rock were blasted, loaded into railcars by the large steam (later electric) shovels and then hauled away to the mill at Magna. This was the beginning of open-cut copper mining in Bingham Canyon on what would become the Utah Copper open-pit mine.

The Utah Copper Company was incorporated in June1903. The Utah Copper mining operation was on the lower part of "The Copper Hill", below the Boston Con's operation. Construction was started

Figure 21 – Startup of Utah Copper in 1906, View Down Canyon
From "Utah Ghost Rails" by Carr & Edwards – This photo is of the same area as the previous photo which was an up-canyon view. Two levels are being cut into the original hillside and the first shovel to get started is on the lower of the two levels. An overpass has been built over the roadway. The railroad on the right is probably the Shay Track. The underground mine, with the covered track that resembles a snow shed, is probably the mine from which ore was first sampled, then mined for metallurgical testing and then to provide 300 tons per day to the Copperton Mill.

in August 1903 of a 300-ton experimental mill in Copperton to test different types of concentrating machinery and to prove that the recovery of copper was sufficient to proceed with larger scale operations. A tunnel was driven into The Hill in September 1903 for the purpose of providing ore for the mill. Ore was at first stoped from sub-level chambers but block caving was soon implemented and provided most of the ore.

In August 1906, two months after the Boston Consolidated, the Utah Copper also started mining by the open-cut method with steam shovels. The ore averaged about two percent copper with 0.15 ounces of silver and 0.015 ounces of gold per ton. Ore was at first hauled to the Copperton mill by Shay engines over the Copper Belt Railroad which was purchased by the Denver and Rio Grande Western Railroad in 1905.

In later years the mills were at Magna and Garfield and the Bingham and Garfield Railroad hauled the ore to these concentrators. Waste rock, or "capping" from both mines was hauled to adjacent gulches where it was dumped.

Both the Boston Con and the Utah Copper increased their open-cut production and reduced their underground production until all their low-grade copper ore came from the open cuts. By open-cut mining the Utah Copper was able to greatly increase the tonnage of ore mined and to reduce the cost of mining to one-fourth of the cost of underground mining.

Open-cut (or open-pit) mining is inherently limited by how steep the slope can be before it will slide downhill. Waste material must be removed from higher up the slope to reduce the slope angle and prevent slides. The Boston Con was near the top of the "Copper Hill" and so had to remove only a small amount of waste rock and its mining of waste rock farther up the hill was not limited by property boundaries. However, the Boston Con only owned the upper portion of the huge orebody which extended downslope into the Utah Copper property. The Utah Copper was at the bottom of the hill and, as it removed waste capping and then underlying ore the slope became steeper and steeper because it couldn't mine material from the Boston Con's property higher up on the mountain. This was a great hindrance to the Utah Copper's operation that would have to be resolved within a few years.

By 1909 the slope angle was becoming critical; the Utah Copper needed to mine away ore and waste from the Boston Con property so that it could continue to mine the ore on its own property. Some sort of arrangement had to be made between the two companies. Overtures to merge the two companies had been made as early as 1905 and in following years but without success. Finally, in December 1909 and January 1910, arrangements were made and a deal concluded by which the value of Boston Consolidated was appraised at 42% of Utah Copper and all shares of the Boston Con were exchanged for 310,000 shares of Utah Copper. With this merger the major obstacle to mining the entire "Copper Hill" was resolved. Coincidentally, Utah Copper now had the largest developed copper orebody in the world. The effect of this merger on the local populace was that, from 1910 to the present, the Utah Copper was by far the largest mining company in Bingham Canyon and had the greatest influence in Copperfield while I was growing up.

The early open cut of the Utah Copper was almost surrounded by the communities of The Boston Con which was on the mountain above it, the Highland Boy, and Carr Fork areas to the north and west, by the main street of Bingham to the east, and by Copperfield and the Upper U.S. to the south. The waste dumps extended only to the nearest gulches that weren't occupied by dwellings.

As the mine expanded, the communities of the Boston Con and the Upper U.S. were the first to be mined out or covered by waste dumps. The open pit continued to expand still farther until it was necessary to mine out the roadway between the main business district of Bingham and that of Upper Bingham which was also known as Copperfield. A one-way, 1 1/4 mile-long vehicular tunnel was driven in 1937-39 to replace this section of roadway. The Bingham-Copperfield vehicular tunnel was opened to traffic February 4, 1939 and was officially dedicated September 30, 1939. After completion of the tunnel the lower levels of the mine expanded to the east so that the main road up the canyon was severed. The mine grew in depth and width year by year, chewing away at the old town slowly but surely. The waste dumps extended farther and farther and covered more of the original hillsides as the mine was deepened.

When the mine became so deep that hauling ore up to the old canyon level became too expensive then a railroad tunnel was driven from the bottom of the pit (elevation 6040) to an exit point halfway down the canyon. A few years later, when the mine was still deeper, the 5840 and, lastly, the 5490 (which exited almost down in Copperton) tunnels were driven.

Finally, in 1958, all the private property in Copperfield was purchased and the town demolished. Shortly after that all the property within Bingham Canyon, which included Highland Boy, Carr Fork, and the Main Street of Bingham was purchased by the mining company and all these communities were either mined out or covered by waste dumps. Even the town of Lark was bought out and then demolished. Only the town of Copperton, at the mouth of the canyon, still survives.

The method of haulage of ore and waste by railcars was replaced by huge dump trucks -- first on the upper waste levels and later on the lower ore levels. Conveyor belts eventually replaced the rail haulage of ore from the mine to the mill. Underground mines around the periphery of the open pit were shut down (or mined out by the open pit or covered by waste dumps) one by one over the years until the Lark Mine (through the Mascotte Tunnel) was the only one left and then it, too was closed. The dimensions of the open-pit mine in 2006 are approximately 2200 feet in elevation from top to bottom and two miles across.

MEMORIAL TO BE HELD HERE

Opening of New Utah Copper Gemmell Memorial Club House to be Opened Monday Evening

The new Gemmell Memorial Club House of the Utah Copper Company will be opened with Memorial services given by Rev. Simpkins of Salt Lake City, Monday evening at 7:30.

The new edifice which has been in the course of construction for the past year, will be one of the finest club rooms in the country, being equipped with everything necessary for the comfort of the workers of this great institution.

Following the memorial services, an inspection of the building will be made, followed by a dance.

The Club House is a memorial to R. C. Gemmell, who died about a year and a half ago, being at the time of his death General Manager of the Utah Copper, and managing director of the entire Utah Copper holdings.

The above article was printed in the October 3, 1924 edition of The Press-Bulletin, Bingham Canyon, Utah.

STEAM SHOVELS AND STEAM ENGINES

Steam shovels were the most significant new machinery to be employed in copper mining by the Utah Copper Company. Seven Marion and one Vulcan steam shovel were being used in the spring of 1909. Some of these shovels had previously been used to dig the Panama Canal. The first steam shovels moved on wheels along railroad tracks so an additional set of tracks for the shovel was laid between the train tracks and the bank of ore or waste that was to be mined. The railroad wheels and tracks can be seen beneath the shovel in the photo below. A large crew of men was required to operate the early shovels.

Figure 22 – Early "Standard" Steam Shovel on Railroad Wheels
From the Gordon Bodily Collection - This shovel, with cables holding the dipper and with no identification on the side, probably came from the Panama Canal. Note the "railroad" tracks along which the shovel moved forward and back. Only three men are shown in this photo but the entire crew included several more. The "shovel runner" ran the overall operation – he moved the shovel forward and back along the track, rotated the boom and moved the stick that the dipper is attached to; his was the highest-paid job on the hill. The most unusual (and dangerous) job on a shovel was that of the "craneman" who rode outside the cab on the "circle" at the base of the boom and pulled a rope to release the ore from the dipper when it was positioned above the ore car. The A-frame at the base of the boom was stationary; the "circle" was right next to it and rotated with the boom. The "craneman" had to be very careful not to be caught between the stationary A-frame and the rotating boom – there were incidents of heads and legs being crushed.

Figure 23 – Gear-Driven Shay Engine of the Copper Belt Railroad About 1907
Used by Permission, Utah State Historical Society, all rights reserved – This engine was working in the open-cut mine about 1907. Shay engines were already being used by the Copper Belt Railroad to haul ore from the underground mines so they continued to be used for a few years in the open-cut mine. However, the Shay engines were slow, the Copper Belt roadbed was quite steep and hazardous and mine production quickly exceeded the capacity of the Copper Belt Railroad. The need for a better railroad was immediately apparent. In 1905 the Denver & Rio Grande Western Railroad began construction of the High Line to replace the Copper Belt Railroad. The High Line eliminated the steep grades and switchbacks of the Copper Belt and allowed use of standard gauge engines. From 1907 to 1911 the D&RGW High Line hauled all the ore from the open-cut mine to the Magna mill. The High Line set new records in ore haulage but still couldn't keep pace with the increasing ore production. The High Line was replaced in 1911 by the Bingham and Garfield Railway. The use of the Copper Belt Railroad and the Shay engines in the open-cut mines of the Boston Con and the Utah Copper was limited to the years of 1906 to about 1908.

EARLY YEARS OF OPEN-CUT MINING

Figure 24 – Steam Shovel and Train On Uppermost Boston Con June 1907
Used by Permission, Utah State Historical Society, all rights reserved - The very top of the "Copper Hill" is being mined for the first time by this steam shovel that is loading waste capping into rail cars. Two steam engines and possibly a second shovel (on right) are shown. The steam engines will push or pull the cars out to the dumps where the waste rock will be tipped over the edge to roll and slide down the hillside. A drill is apparently drilling vertical blast holes from the small cut above the shovel. The hillside above this level is relatively undisturbed but soon will be removed as the Boston Con (which will become part of the Utah Copper in 1910) strips waste rock from higher levels to allow the mining of ore from levels farther down the hillside.

Figure 25 – Steam Shovel and Engines on Steep Hillside of the Boston Con May 1, 1907
Used by Permission, Utah State Historical Society, all rights reserved - The Boston Consolidated started mining with steam shovels high up on "The Copper Hill" in 1906 several months before the Utah Copper started similar operations near the bottom of the hill. This photo shows new levels being dug into the steep mountainside. The operations of the two companies were separated by native hillside until about 1910 when the two companies merged and the restriction to mining the hillside above the Utah Copper was eliminated.

Skinks are reptiles similar to a lizard but with a longer body similar to a snake. They often have long blue tails that break off and twist around on the ground when you try to catch them. There definitely were skinks in the Bingham - Copperfield area but only a few people saw any. Near Black Rock above Telegraph and in Butterfield Canyon are two locations where sightings have been reported.
Did you see any skinks in the area of Copperfield or Bingham? If so, where at?

Figure 26 – Loading Ore at Utah Copper Mine About 1908
From "Bingham, The Greatest Copper Camp In The World" 1909 – Two steam engines are shown – one train is loading while other waits its turn.

Figure 27 – Loading Waste at the Utah Copper Mine About 1908
From "Bingham, The Greatest Copper Camp In The World" 1909 – The Standard shovel, the steam engine and the wooden railcars are all of the earliest vintage.

Figure 28 – Steam Engine and Steam Shovel, Early Open-Cut Mine About 1906
From the Gordon Bodily Collection - Waste rock or "capping" is being removed in this operation which is high on the West Side of the open cut mine; it is probably part of the Boston Consolidated operation. The hillside above the shovel still has brush on it and the rock beneath would have been leached of the copper it originally contained. The cars that the engine is pulling are for hauling waste rather than ore. The low height of the bank also indicates that this was an upper level that was just being developed. Note the short segment of railroad beneath and behind the shovel; in this instance more track would be laid as the movement of the shovel required it. Unused ties are in the foreground for use as needed. The steam engine appears to be a narrow-gauge Shay engine that is utilizing the outermost two of three rails that comprised a narrow-gauge/standard gauge track. Other engines and cars of standard gauge width would be able to use the two farthest-apart rails. The shovel could be one of the early "Standards" from the Panama Canal.

Does anybody know the details of the fake stick of dynamite that was thrown into the iron shack and the guys inside all scrambled out through the windows?

Figure 29 – Dinkey and Cars On Trestle Across Bottom of Utah Copper Mine 1908
From the Gordon Bodily Collection - This trestle allowed waste or "capping" that was mined from the up-canyon limit of the mine to be hauled down the canyon and around the hill to a temporary dump toward Carr Fork while at the same time ore was being mined from the developing "pit" below and behind the bridge. Two sets of railroad tracks pass beneath the bridge; ore was hauled out of the mine on these tracks and then down the canyon to the mill.

What was the wooden trestle in the bottom of the pit called?

Figure 30 – Panorama of the Copper Hill and Two Open-Cut Copper Mines in 1908
Photo Courtesy of Kennecott Utah Copper – Mine levels of the Utah Copper are in the bottom half of the photo; ore is being mined from the right and waste is being dumped to the left of the photo. Levels of the Boston Con are in the upper right. Copper Center Gulch runs from the upper center of the photo to the middle left edge. Upper Bingham, or Copperfield as it came to be known, is in the bottom of the canyon near the left side. The East-Line Extension Bridge is to the left. The longer, higher, straight E-Line Bridge had not yet been built when this photo was taken (It would be built in 1912). The Copperfield that most of us knew was up-canyon from this bridge. When we were growing up in Copperfield we never heard of Copper Center Gulch, the mouth of which was about where the Niagara Tunnel and U.S. Hotel were located. Essentially all of Copper Center Gulch above the Jap Camp-to-Greek Camp road was filled with waste rock within a couple years after the open-cut mine was started.

Figure 30 (Continued) – The Copper Hill and Two Open-cut Copper Mines in 1908
Photo Courtesy of Kennecott Utah Copper - The main part of the Utah Copper orebody is in the left half of the photo from the canyon bottom to halfway up the hill. The Boston Con is at the top of the hill. There is an unmined section of the mountain between the upper workings of the Utah Copper and the lower levels of the Boston Con; it will be quickly mined away after the 1910 purchase of the Boston Con by the Utah Copper. The shovels in and near the bottom of the mine are working beneath extremely high banks which the shovelrunners were very leery of. In fact all of the banks appear to be higher than the 50-foot height that was later adopted as the standard height for all levels. Carr Fork and the main part of the city of Bingham are in the far right side of the photo.

ORES SHIPPED FROM THE BINGHAM DISTRICT THIS WEEK

U. S. Mining Co.	39 cars
Utah Apex Mining Co.	33 cars
Bingham Mines Co.	8 cars
Mont. Bingham Co.	9 cars

The above article was printed in the August 1, 1922 edition of The Press-Bulletin, Bingham Canyon, Utah.

Figure 31 – Steam Shovels Working Back Into "The Hill" in 1909
From the Gordon Bodily Collection - This photo shows the early years of the development of the open-cut mine and its encroachment upon Upper Bingham. Waste rock and "capping" are being dumped to the left (south) in Copper Center Gulch and the north side of Galena Gulch. The houses and other buildings are threatened both by being mined away and by being covered by waste dumps -- this proved to be the ultimate fate of all the buildings in Bingham Canyon including the town of Bingham and all the adjoining smaller communities including Copperfield. In 1905 "the existence of the railroad, county road, and dwellings in the gulch" had been pointed out as an "unfavorable feature" which might hinder the use of steam shovels (Old Reliable, p. 53). In actuality the mine was able to expand as necessary until the entire town was gone.

STEVE HAYS MINING PROPERTY SOLD

The Steve Hays' mining property, located east of Copperfield and adjoining the Bingham Mines and Fortuna properties were sold the past week to mining men of Salt Lake City. The land sold consists of sixty acres and includes the south end of the Life Guard claim and all of the Claret, Butter, Agie, Weasel Extension and Old Trail, Nos. 1 and 2 claims. The sum paid for the properties was $27,000.

The above article was printed in the March 22, 1924 edition of The Press-Bulletin, Bingham Canyon, Utah.

Figure 32 – The Utah Copper Mine in 1909
Used by Permission, Utah State Historical Society, all rights reserved - This photo is looking up-canyon toward Upper Bingham (Copperfield). One or more dropcuts have been made since 1906 so that a pit is beginning to form. Coal smoke obscures the scene; electrification of the operation is far in the future. The area being mined is part of the higher grade orebody which extends still farther into the hill to the right. The company was surely "making money" from the work shown in this photo. These early operations were very labor intensive compared to modern methods and the machinery that is now being used but they were still very profitable. The trestle on the left is the same as in the 1908 photo of the "Dinkey and Cars On Trestle 1908". Note that the railroad under the bridge has been rerouted and the pit in the foreground has been deepened since 1908.

ORES SHIPPED FROM THE BINGHAM DISTRICT THIS WEEK

U. S. Mining Co.	42 cars
Utah Apex Mining Co.	22 cars
Bingham Mines Co.	11 cars
Mont. Bingham Co.	9 cars

The above article was printed in the July 29, 1922 edition of The Press-Bulletin, Bingham Canyon, Utah. Used by Permission, Utah State Historical Society, all rights reserved.

View of Upper Crossing of Utah Copper Co.'s Waste Track over Property of U. S. Mining Co. Material too low in value to treat is taken over these these waste lines to dump ground.

Figure 33 - Upper Crossing (H-Level Wooden Bridge) About 1908
From the Gordon Bodily Collection - This wooden bridge can be seen in a couple of later photos of Copperfield; this is the earliest. The wooden bridge was replaced in later years by the concrete overpass that we were familiar with in the 1930s. The long snowshed-like structure above the bridge is the Holden Tramway (gravity and mule-powered) with a covering over the tracks. An aerial tramway "control station" is in the left foreground; it would have to be part of the tramway from the U.S. Mine (Jordan and Galena mines) in Galena Gulch to Lower Bingham that operated from about 1902 to 1910. Cables can be seen attached to it but no buckets are definitely visible. The small wooden trestle or overpass in the center foreground and the grade behind it could be part of the D&RGW Railroad going up to the U.S.'s mines in Galena Gulch; at least some ore was hauled on it rather than the aerial tramway after about 1907.

RAILROAD LINES

There were half a dozen or more railroads and tramways built in Bingham Canyon after mineral was found in Galena Gulch in 1863. The railroads included both narrow-gauge and standard gauge. The earlier railroads that went up to the mines, such as the Copper Belt, were narrow gauge and had steep grades and sharp curves which limited the engines that could be used and their capacity for hauling ore. There were several tramways, most of them aerial. Highland Boy had three aerial tramways and there was one in the Copperfield area that transported ore from the mines in Galena Gulch and Telegraph down to Lower Bingham. The Holden Tramway was at first a 20-inch narrow-gauge mule and gravity tramway that was converted in 1899 - 1901 to Shay engines and in 1902 became part of the Copper Belt Railway. The Copper Belt was supplanted by the High Line of the Rio Grande Western in 1905. The High Line greatly increased the tonnage of ore that was hauled from the open-cut mines, setting new records of as much as 6000 tons of ore daily but it was still not able to keep up with the increased mine production. The High Line was replaced by the Bingham and Garfield Railway in 1911.

Figure 34 – Horseshoe Bend on High Line of Rio Grande Western Railroad About 1908
From "Bingham, The Greatest Copper Camp In The World" 1909 – The Horseshoe Bend was about halfway between Lark and Copperton. The trestles, tracks and ties had been removed many years before the 1940s. This was a favorite area for hunting cottontail rabbits. We would start early in the morning, walk all the way from Copperfield, hunting up and down the brush-covered gulleys along the way, and then walk back home late in the afternoon or evening.

Looking down Bingham Canyon. Copper Belt R.R. just above buildings. R. G. W. High Line above, and two lines of Utah Copper Co.

Figure 35 – Railroad Lines in Bingham About 1909

From "Bingham, The Greatest Copper Camp In The World" 1909 - This photo was probably taken in 1909 but the historical records are somewhat confused. The abandoned Shay Track of the Copper Belt Railroad is just above the buildings; it originally continued way up the canyon to the U.S. and around T he Copper Hill to the Highland Boy area. It was also evident just above the buildings on the east side of Main Street in Copperfield. The High Line of the Rio Grande Western, with the Cuprum Yards, is the main rail line halfway up the mountain. The High Line replaced the Copper Belt Railroad. A-Level in the mine was the same level as the Cuprum Yards of the D&RGW. The two rail lines above the High Line are Utah Copper dumplines, probably C-Level and E-Level over which waste rock was hauled farther down the canyon. If one followed E-Level a mile farther up the canyon in the mine, you would come to the East Line Extension Bridge in Copperfield (Replaced by the E-Line Bridge in 1912). The main Bingham Canyon extends up-canyon to the lower right of this photo; Carr Fork extends up-canyon to the lower left.

EARLY YEARS OF COPPERFIELD

The early years of Copperfield are shown by the photos and are described by their captions. Most of us weren't yet living in Copperfield when these photos were taken. Many of the features, such as the Holden Tramway, the aerial tramway from the U.S., the Copper Belt Railroad and the Upper Crossing (H-Level Wooden Bridge) were removed before the 1940s so we were never able to see them.

Utah Copper Property as Seen From Upper Bingham—East Line Extension Bridge.

Figure 36 – Utah Copper, Upper Bingham, East-Line Extension Bridge About 1907
From "Bingham, The Greatest Copper Camp In The World" 1909 – This photo is looking down the canyon from the Ohio Copper area about a year after the Utah Copper open-cut mine was started. Most of the levels are still being established. This curved bridge was replaced in 1912 at essentially the same location by the E-Line Bridge that was higher, longer, and straight. The Shay Track of the Copper Belt Railroad is the railroad track in the bottom of the photo near the center; it still has the tracks on it and appears to have still been in use. Note the tents on the hillside on the right of the photo.

Figure 37 – Lower Copperfield, Niagara Tunnel, East-Line Extension Bridge About 1907
From the Gordon Bodily Collection – Copperfield was known as Upper Bingham at this time. The Niagara Tunnel and the East-Line Extension Bridge are two geographic markers that were relatively constant for many years. The smokestacks of the air compressor/heating plant and some of the dumps of the Niagara Tunnel remained; the U.S. Hotel and orebins were built adjacent to them several years later. This curved bridge was replaced, in essentially the same location, in 1912 by the higher, longer, straight E-Line Bridge which remained until about 1941. At the extreme left of the photo is the Shay Track, complete with ties and rails, that we knew, in the 1940s and 50s as the gentle pathway (without ties or rails) behind the houses on Main Street and just above the entrance of the vehicular tunnel that we called "The Shady Track". The Holden Tramway is near the upper left corner of this photo – the track is covered over by a wooden snowshed-like protective cover. Fox's Switch and H-Level have not been built yet. This photo presents a good view of Copper Center Gulch which runs from the upper right of the photo down to its mouth by the smoke stacks of the compressor for the Niagara Tunnel. The levels for hauling waste into this lower part of Copper Center Gulch haven't been built yet so no waste rock has been dumped into it. The Copperfield (Upper Bingham) Grade School that opened in 1909 has not been built yet.

Figure 38 – Copperfield Grade School, Aerial Tramway and Utah Copper in 1910
Used by Permission, Utah State Historical Society, all rights reserved - This is the first Copperfield (Upper Bingham) School; it was constructed in 1908, opened in 1909 and burned down in 1918. The furnace and bathrooms were inside the schoolhouse. The large chimney for the furnace is on the front of the school which isn't visible in this photo. Ore buckets and cables of the U.S. Company's aerial tramway can be seen in the upper left of the photo; this tramway carried the ore from the U.S. Company's Jordan and Galena mines in Galena Gulch down to the Denver and Rio Grande Western depot in lower Bingham. The dump, ore chutes, smoke stack for the compressor, and buildings of the U.S. Company's Niagara Tunnel are just above and to the right of the schoolhouse. The Utah Copper open-cut mine is halfway up on the right side of the photo; the Boston Con workings (also Utah Copper at this time) are on the top of the mountain in the center of the photo. Copper Center Gulch is the large gulch to the left of the Boston Con open-cut; it emptied into the main canyon just about where the Niagara Tunnel surface plant is located. The Utah Copper began hauling waste rock and dumping it to fill in Copper Center Gulch immediately after the open-cut mine was started. By the 1920s very few traces of this gulch were apparent and younger residents didn't know it ever existed.

Figure 39 – Early Utah Copper and Copperfield With Aerial Tramway About 1911
From the Gordon Bodily Collection - The top of one support tower of the aerial tramway is in the center at the very bottom of the photo. Two or more buckets can be seen hanging from the cables down-line from the control station. The aerial tramway was that of the U.S. Company that transported ore down from the Jordan and Galena mines at the head of Galena Gulch. Just above the tramway tower are two big smokestacks of the air compressor/heating plant of the U.S. Company's Niagara Tunnel. Greek Camp and the U.S. Hotel have not yet been built. On the right edge of the photo is the Shay Track of the Copper Belt Railroad (D&RGW at this time); it still has tracks on it and is apparently still in use. The Copper Belt Railroad was purchased by the Denver and Rio Grande Western Railroad in 1905. Farther down the canyon, on the right side, is the East-Line Extension Bridge of the Utah Copper. The wide area in the canyon where the Niagara dump, orebins and compressor plant and other buildings are located is the mouth of Copper Center Gulch. Ore is being mined by the Utah Copper from the central area of the photo and waste is being dumped to the left of the photo in the upper part of Copper Center Gulch or hauled around to the East Side dumps. Three historically significant methods of ore haulage are shown here – the narrow gauge Shay Track, the aerial tramway and Utah Copper's standard gauge railway system. The Boston Con and the Utah Copper open-cut mines haven't merged yet; there is unmined hillside between the two. The date of the photo is about 1911.

Figure 40 – Main St. With Greek Camp, Jap Camp, and Three Bridges About 1916
Photo Courtesy of Kennecott Utah Copper - The curved East-Line Extension Bridge is near the bottom of the photo but is no longer being used; it wasn't torn down for many years. The newer, higher, straight E-Line Bridge (built in 1912) crossed over the west abutment of the older bridge. The Upper Crossing wooden bridge that crossed the canyon at H-Level is halfway up on the left edge of the photo; it was replaced in the 1920s by the concrete overpass that most of us remember. Greek Camp is slightly to the right of center; Jap Camp is to the left of Greek Camp just below H-Level. The Niagara Tunnel of the U.S. Company is below Jap Camp with the smokestacks of the compressor/heating plant, dumps of the tunnel and the ore-dumping structure. The U.S. Hotel has not yet been built. Copper Center Gulch, which passes between Greek Camp and Jap Camp, has been filled in with waste rock. The steam shovels and engines of the Utah Copper haven't been electrified yet so no overhead electrical wires are present above the tracks of the levels or of the bridge.

Figure 41 – Three Copperfield Bridges 14Oct1916 View Up Canyon
Used by Permission, Utah State Historical Society, all rights reserved – The E-Line Bridge is on the extreme right of the photo. The East Line Extension Bridge is in the center of the photo – the left (east) side of this bridge has been dismantled. The Shay Track of the Copper Belt Railroad went under the old bridge and then farther up and down the canyon. The houses and other buildings between the two bridges are shown. Some Copperfieldites (Duggie Gonzales) lived almost under the E-Line Bridge and can remember several of these buildings. The Ohio Copper shaft and associated buildings are in the upper left of the photo. The Upper Crossing (H-Level Wooden Bridge) can be seen farther up the canyon. Main Street, Greek Camp and Jap Camp are evident. The Copperfield Grade School had already been built and is in the photo but, due to smoke and fading of the negative, is very difficult to see.

When was the East Line Extension Bridge demolished?

Figure 42 – Upper Copperfield Before Main Street Houses About 1921
Photo Courtesy Of Kennecott Utah Copper - The second Copperfield Grade School (the first one burned down in 1918) is on the right. It was built and opened in 1919. It does not yet have the auditorium/school rooms addition on the back (it will be built in 1926). Buildings immediately below the school house and also across from and above the school were either burned down in the July 1918 fire or have been demolished in preparation for the construction of new Utah Copper company homes on Main Street. The houses on Main Street will be present in 1924 photos. The road bed of the Shay Track runs just behind the school building but the tracks have been removed. Farther down the canyon the most recognizable structures are the two wooden bridges, the nearer, smaller, curved one is the East-Line Extension Bridge which is out of service; the farther, larger, straight one is the E-Line Bridge that was built in 1912. Jap Camp is on the far left of the photo; below it is Fox's Switch. Two leaser's ore cars are on the side track of Fox's Switch. The U.S. Hotel has not yet been built. A ramshackle collection of wooden houses occupies the hillside where the 2-story duplexes of Terrace Heights will be built from 1925 to 1927. In the right foreground of the photo are several railroad tracks and switches and the end of a trestle; these were probably part of the D&RGW High Line going up to the U.S.'s mines in Galena Gulch. Note that between the E-Line Bridge and H-Level on the East Side (right of photo) there is considerable of the old hillside still in place; this can be used for dating photos.

Figure 43 – Down Canyon View of Copperfield in 1934. Upper Main St to E-Line Bridge
From the Gordon Bodily Collection - The Utah Copper open-pit mine is in the background of the photo. Greek Camp is in the upper left, Jap Camp in the left center, the U.S. Company's Niagara Tunnel, ore bins, compressor and hotel are midway up the photo to the right of Jap Camp and Greek Camp. The Utah Copper company-built houses are in line along Main Street with the Grade School on the right of the street. Farther down the street, near the center of the photo is the U.S. Hotel. The E-Line Bridge is farthest down the canyon above the open pit. The two-story duplexes of the Lower Terrace Heights are on the hillside above the Grade School on the right edge of the photo. Buildings of the Ohio Copper Company are at the same elevation as the Lower Terrace Heights farther up on the photo toward the upper right corner. The extension to the back of the school building can be seen – it was constructed in 1926. The Vehicular Tunnel has not been built yet – it will be built across the street from the U.S. Hotel in 1937-39. The abandoned Shay Track is on the right (east) side of the canyon just above the Grade School and the houses that line Main Street; it goes under the E-Line Bridge but a side track goes up to E-Level by the east side of the bridge.

When was the U.S. Hotel built?

Figure 44 – Copperfield Looking Up Main St With Greek Camp, Jap Camp 1927-1937
From the Gordon Bodily Collection - In this photo the U.S. Hotel has been built and the Utah Copper operation has been electrified. An electric locomotive is on the switchback going up from Fox's Switch just below Greek Camp; electric locomotives were first brought into the mine in 1927. The roof of the Copperfield Grade School may be visible halfway up on the left side of the photo. The concrete overpass for H-Level has replaced the wooden Upper Crossing. The ore bins for the mine that went through the mountain by Carter's house is above the roof of the school. The top of the chute for the ore bins is even with the bed of the Holden Tramway which goes slightly uphill to Telegraph. The road to Dinkeyville slopes from the top of the ore bins down to the main road above the concrete overpass. The bottom of the ore bins is on H-Level. Buildings and the dump of the Ohio Copper are on the left side of the photo. The vehicular tunnel does not appear to be present so this photo was taken between 1926 and 1937 when construction on the tunnel was started. No features are evident that could define the date of the photo any closer.

FINN DEMONSTRATES HIS STRENGTH

The home of City Clerk Quinn in Carr Fork is enclosed by an ornamental iron fence. On Friday night Gus Dahl whilst returning home spied the different projections and decided to demonstrate his ability as a strong man to his friends, and twisted the projections in all manner of shapes. He was brought before City Judge Dudley and charged with intoxication and destruction of property, on Wednesday morning, and was sentenced to fifty days or a $50 fine. He is still in gaol.

The above article was printed in the July 8, 1922 edition of The Press-Bulletin, Bingham Canyon, Utah.

Figure 45 – Miner's Merc., Lower Main Street in About 1934
Photo Courtesy of Shirl and Isabell Scroggin - The Independent Grocery, Miner's Mercantile and the adjoining building that was occupied by Mike the Barber, Joe Berger's Combination Bar, the Pan-Hellenic Grocery and the Byrnes Apartments were mainstays for decades. The homes and other buildings behind and below these three were torn down shortly after the E-Line Bridge was demolished. The shadow of the E-Line Bridge is present on the left edge of the photo. The old Shay Track is halfway up the photo on the right side – just behind the houses on Main Street. The Shay Track had been abandoned long before but its roadbed went under the E-Line Bridge and its sidetrack merged with E-Level near the left edge of the photo. The Ohio Copper area is in the upper right of the photo. Mine levels on the East Side extend higher than H-Level. The date of this photo, judging from features of the mine and the town could be from 1930 to 1940. The automobiles on the street, however, appear to be about 1934 models so the photo was probably taken in about 1934.

Figure 46 – Upper Main St., School, Start of One-Story Terrace Heights in 1924
Photo Courtesy Of Kennecott Utah Copper – The Company houses along Main Street have been built. The one-story duplexes of Terrace Heights are in place but might still be under construction. The two-story duplexes of Terrace Heights have not been built yet; they will be built in 1925-1927. The area that the two-story duplexes will occupy still has an array of wooden houses and shacks clinging to the hillside. The roadbed of the old Shay Track extends from the left edge to the right edge about halfway up the photo and passes behind the school house. The Copperfield Grade School is in the center of the photo. The auditorium has not been added to the back of the school – it will be built in 1926-1927. H-Level is in the upper part of the photo, extending all the way across it; some waste cars are sitting on the siding there. No other levels are above H-Level on the East Side but development of I-Level might be proceeding on the left of the photo. Dinkeyville is the assortment of houses above H-Level. The tracks of Fox's Switch are in the left lower corner. The site of the U.S. Hotel isn't visible.

Figure 47 – Copperfield and Utah Copper Mine Looking NW About 1937 to 1940

From the Gordon Bodily Collection - This photo shows most of the features of Copperfield that many of us remember. The E-Line Bridge is on the right edge of the photo; it was built in 1912 and demolished about 1941. The main street of Copperfield extends up the canyon from the bridge to the lower left of the photo. The U.S. Hotel and the ore bins, dump, and smokestacks of the Niagara Tunnel are halfway up Main Street. The Circle is a short distance up the street from the hotel. Some of the Company houses along Main Street are evident. The Upper Terrace two-story duplexes are on the right of the photo; they were built in 1925-1927. An edge of Dinkeyville is in the lower right corner of the photo. A train of loaded ore cars is coming down the hill toward Fox's Switch that is in the lower left of the photo. The Niagara Tunnel, with stacks of lumber, is to the left of the loaded ore cars. Jap Camp is the group of long single-story wooden buildings on the left of the photo above Fox's Switch. Greek Camp is above the Niagara Tunnel this side of the open-pit mine. The edge of the pit has been expanded to very near the E-Line Bridge and is approaching Greek Camp. Mine dumps are extending far down the canyon on the West Side. Telegraph and the U.S. aren't shown in this photo; they would be to the left. The location of the Bingham-Copperfield vehicular tunnel can't be seen but it could be present and would help to more closely date this photo. A guesstimate is that this photo was taken between 1937 and 1940.

Figure 48 – Main Street Copperfield With Structure Above Roadway in 1937

Photo From Bingham High School Alumni Collection; Scan by Rebecca Reid – This view is part of a larger photo. It shows the Main Street from the school house down to the E-Line Bridge. Features include the school house in the lower right corner, company houses on both sides of the street, the U.S. Hotel is the large 4-story building on the left, the business district is farther down the street, and the E-Line Bridge at the top of the photo with the open-pit mine behind it. The old Shay Track is above the houses on the right. The Niagara Tunnel dump is behind the U.S. Hotel on the left. The engine in the upper left is pulling the linecar that is used to work on the overhead electrical lines.

This photo shows a structure that crosses above the roadway just down the street from the U.S. Hotel. Its purpose is unknown to the author – information from readers will be welcomed.

THE MINE CHEWS AWAY THE ROAD

The open-cut mine (which became the open pit as it was deepened) had, almost from its beginning, been threatening the roadway between the main part of Bingham and Upper Bingham, or Copperfield. The tenuous existence of the roadway ended shortly after September 1939 when the vehicular tunnel was dedicated.

Figure 49 – Open Pit With Road Along East Side In 1926
From the Gordon Bodily Collection - The roadway from Bingham to Copperfield can be seen near the bottom of the canyon on the East Side (left) of the mine. The smaller, curved East-Line Extension Bridge with the higher, longer, straight E-Line Bridge right next to it are in the lower part of Copperfield. The East-Line Extension Bridge was built in 1907 and was no longer being used; the E-Line Bridge, which replaced it, was built in 1912. The U.S. Hotel is at the left edge of the photo. The waste dumps have filled in Copper Center Gulch and a large part of Galena Gulch

(the U.S.). On the East Side of the mine H-Level is the highest. In the bottom center of the photo A-Level is the lowest level in the mine and will remain so for several years. The area of the "pit" that is being formed over the center of the orebody is only down to D or E-Level and the width of the "pit" is very small. The lack of smoke indicates most, if not all, of the steam shovels and some steam engines had been converted or replaced by electric models. This photo was taken May 7, 1926.

Figure 50 – Bingham-to-Copperfield Road Along East Edge of Pit In 1926
From the Gordon Bodily Collection - The pit is being pushed on the south and East Sides right up next to the roadway. This photo shows two of the shelters alongside the road to protect people and autos from flying rocks during blasting. These open-sided but roofed shelters were large enough to accommodate one or two automobiles. The length of time between the sounding of the warning whistle and the actual blast was somewhat uncertain so kids who were in a hurry would, after the warning whistle, run from one shelter to the next rather than just wait beneath the nearest shelter. "Whistle Joe's" shack is across the track. He blew the whistle that signaled work to start, lunch break, quitting time etc. This photo was taken in 1926.

Figure 51 – Utah Copper Mine With Road Thru Middle About 1932
From the Gordon Bodily Collection - The E-Line Bridge is still in place. The shovels and engines have been electrified. The East Side levels extend up to about L-Level. The levels in the pit aren't pushing against the E-Line Bridge yet. The area of the "pit" that is being formed over the center of the orebody is down to about C-Level. It has been mined back considerably to the west and to the north so that the width of the "pit" is considerably greater than in 1926. A guesstimate is that this photo was taken between 1929 and 1934.

KENNECOTT HOLDS THE CONTROL OF UTAH COPPER MINE

Kennecott, which today owns such a big block of Utah Copper as to be declared to now practically control the Bingham proposition, is now more or less of interest to local traders. The Kennecott Copper Company's production in June, totaling 4,044,000 pounds, showed an increase of 640,000 pounds over May. Braden, Kennecott's big South American subsidiary turned out 8,292,000 a new high record, proof conclusive that it is coming up to expectations. Kennecott directors have repeatedly emphasized the statement that if anything ever occurred to curtail production in Alaska the corporation could go right along with Braden and pay dividends from its future output alone.

The above article was printed in the July 19, 1918 edition of The Press-Bulletin, Bingham Canyon, Utah.

Figure 52 – Road Below E-Line Bridge Being Mined Out by Shovel in About 1940
Photo Courtesy of Eugene H. Halverson - The roadway between the Town of Bingham and the community of Copperfield is being severed by the shovel on the left. This was done after the Bingham-Copperfield vehicular tunnel was opened to traffic in 1939. The E-Line Bridge was apparently still in use. This photo was taken in about 1940.

COPPERFIELD DRUG STORE ADDS PRESCRIPTION DEPT.

The United Drug Store of Copperfield has during the past week made many new improvements, the most noteworthy being the addition of a prescription department, with a registered pharmacist in charge.

Mr. James Jemas, manager of the drug store stated that a complete line of drugs and sundries had been installed, and that the United Drug Store is now as well equipped to care for the needs of the Copperfield people as any store in the state.

Perfumes, and all beauty requisites that might please milady are now in stock, with a complete line of Nyal Remedies, and a soda Fountain service second to none.

The above article was printed in the October 10, 1921 (or 1924) edition of The Press-Bulletin, Bingham Canyon, Utah.

CONSTRUCTION OF THE BINGHAM-COPPERFIELD VEHICULAR TUNNEL

When the slope on the East Side of the pit below the roadway became so steep that it began to hamper mining of ore from the pit then it became necessary to mine out the road. A single-lane vehicular tunnel was built to provide access between Bingham and Copperfield prior to mining away the road. Construction of the tunnel was started in March of 1937 and it was first opened to traffic February 4, 1939. The official dedication of the tunnel was at the first Galena Days on September 30, 1939. The tunnel was a mile and a quarter long and had a steady uphill grade from Bingham to Copperfield of 6.4 percent. The tunnel remained open for twenty-seven years; it was permanently closed to traffic on January 24, 1966.

Figure 53 – Preparation for Construction of Lower Portal of Tunnel in 1937
Photo Courtesy Of Kennecott Utah Copper - Buildings are being demolished from just below the Bingham Hotel down to the City Hall. The roadbed of the abandoned Copper Belt Railroad (Shay Track) is just above the buildings. Some of the landmark buildings shown here are the City Hall (the flat-roofed two-story building with the white trim on the front) on the far side of the street in the left of the photo and the Bingham Hotel on the far side of the street above the demolished buildings. The Bingham Merc is the large building in the left foreground with the triangular-shaped flat roof. The Telephone Office is the two-story flat-roofed building straight across from those that have been demolished. Up the street from the Telephone Office are two three-story apartment buildings.

Figure 54 – Construction of Lower Portal Bingham-Copperfield Tunnel In 1937
From the Gordon Bodily Collection - Construction of the vehicular tunnel started in March of 1937; it was opened to traffic February 4, 1939. Work was performed from both ends of the tunnel simultaneously and the two underground headings met very closely. The tramway on the upper left side of the photo carried construction materials down to the tunnel entrance from the abandoned Shay Track above.

Figure 55 - Lower Portal of Tunnel After Completion
Courtesy Of Kennecott Utah Copper - This photo was taken several years after the 1939 completion of the tunnel. All the buildings between the Bingham Hotel and City Hall were demolished, the tunnel was completed and a new frame building (equipment garage) was built just below the tunnel. The temporary tramway from the abandoned Shay Track (Copper Belt Railroad bed) down to the tunnel entrance was removed and the abandoned railroad bed has been replaced by Kennecott's track running down the canyon from the 6190 Yard and level.

RAILROAD WORKMEN ROLL SNOW BALLS INTO TOWN

For some distance high up on the mountain side the Denver & Rio Grande railroad parallels the Main street of Bingham, and for the past week it has had a force of men shoveling the snow from the track. When the snow is thrown over it starts tumbling down the long, steep incline towards town. As it rolls along numerous snowballs are formed which grow in size and increase their velocity and finally strike the houses on Main street with great force, causing damage and inconvenience to the residents.

It is understood that a number of people are making complaints against the railroad rolling its big snowballs down the mountain side into town. Of course the snow has to be moved from the track, but it is thought that some plan might be devised where the residents of the town would not be disturbed or endangered.

The above article was printed in the February 22, 1918 edition of The Press-Bulletin, Bingham Canyon, Utah.

Figure 56 – Upper Portal of Vehicular Tunnel, Start of Construction 1937
Photo Courtesy of Shirl and Isabell Scroggin – Demolition of the houses at the site of the tunnel portal was the first step toward construction. The two company-constructed houses on the right side of the street below the tunnel site remained in place for the next twenty one years. The Shay Track is immediately above the construction site.

COFFEE HOUSE GAMBLERS ARE ARRESTED

Among the arrests made this week were five alleged gamblers who were found in the Canton Coffee House at Copperfield. The personnel of this party is Mike Mangas, Sam Alex, John Shinnis, John Kalis and John Topalos.

The arrests were made by Deputies Kenner, Fullmer and Jones. The men it appears were conducting their little game behind bolted doors when the officers arrived on the scene and asked for admission. They were refused admission and to effect an entrance broke through the door and captured the offenders.

The men were brought before Judge John C. Green and were admitted to bail in sums ranging from $25 to $100. They will be tried next Wednesday.

The above article was printed in the May 31, 1918 edition of The Press-Bulletin, Bingham Canyon, Utah.

Figure 57 – Construction of Upper Portal of Bingham-Copperfield Tunnel In 1937-38
Photo Courtesy of Shirl and Isabell Scroggin - The Bingham-Copperfield vehicular tunnel was constructed from both the upper and lower portals at the same time. The surveying and control of the construction was very precise so that the two underground headings met with practically no error. The site was fenced off from the main roadway and construction commenced. Some of the equipment used to drive the tunnel is shown. Compressed air, electricity, water pumps and ventilation were necessary to support the equipment that was used for the drilling, blasting, loading and hauling of the broken rock. The old Shay Track and a fire hydrant shed are just above the tunnel.

JACKLING YACHT IS BLOWN UP BY HER ENGINEER

Colonel D. C. Jackling's yacht Cyprus, built in Seattle, was sunk recently, two days out Halifax, after having been sold to the Russian government.

After the sale a Russian crew and officers took charge of the vessel. The engineers did not understand oil-burning boilers, and one of these exploded at sea. The vessel went down almost immediately and three lives were lost.

This news was received in San Francisco on Saturday, says the San Francisco Examiner, in a letter from Captain George Lewis, formerly in command of the yacht, to Captain J. Swanson.

Captain Lewis is now dockmaster for the Morse Drydock company of Jersey City, N.J.

The above article was printed in the January 11, 1918 edition of The Press-Bulletin, Bingham Canyon, Utah.

Figure 58 – Inside the Tunnel, Door Down, Fan Visible, Copperfield Portal About 1939
Courtesy of Kennecott Utah Copper - When the door was down the fan would ordinarily be operating to draw the automobile exhaust fumes out of the tunnel. The raised walkway and guard rail are on the right with the sewer and storm drain underneath the walkway below the level of the road. Overhead lights are spaced along the ceiling above the walkway. The panel for the "electric eye" is in the wall to the left of the vehicle. The revolving door (for pedestrian access while the main door is closed) is to the right of the walkway by the vehicle.

GIANT CAP BLOWS OFF YOUNG GIRL'S HAND

Alpha Pope, 10 years of age, daughter of Mr. and Mrs. George Pope of Copperfield, suffered the loss of her left hand on Tuesday evening by amputation.

The little girl's cousin brought a giant blasting cap to the Pope home and not being aware of its danger applied a lighted match to it with the result that her left hand was blown almost to pieces, besides lacerating her face, neck and chest. She was immediately moved to the Dr. F. E. Straup hospital where she was attended by Dr. J. Flynn who saw the impracticability of trying to save the injured member and it was amputated. On going to press we learn with pleasure the little patient is recovering nicely.

The above article was printed in the July 1, 1922 edition of The Press-Bulletin, Bingham Canyon, Utah.

Figure 59 – Dedication of Vehicular Tunnel September 30, 1939, Copperfield Portal
Photo Courtesy of Shirl and Isabell Scroggin - The Dedication of the Vehicular Tunnel between Bingham and Copperfield was held on Saturday, September 30, 1939 at 11:00 A.M.; it was also the first Galena Days celebration. This photo shows dignitaries and a group of people on the left. The tunnel had actually opened for traffic February 4, 1939. The tunnel entrance and nearby buildings, as they are here, changed very little if any until Copperfield was demolished in 1958. The tunnel remained open for traffic by workers until 1966. It had a useful life of twenty-seven years.

The upper part of the E-Line Bridge is just below the center of the photo. The roadway between Copperfield and the main part of Bingham went under the bridge and the road was used for over 70 years from its inception in the 1860s until the vehicular tunnel was completed in 1939. The opening of the tunnel for traffic and its dedication were major events in the history of Copperfield. After the dedication all vehicular and pedestrian traffic to and from Bingham had to pass through the vehicular tunnel. The tunnel became the "lifeline" that connected Copperfield to the rest of Bingham and to the valley. The dedication signaled a significant change in the living habits of the people. The tunnel also became part of the tourist attraction of Bingham along with the open-pit copper mine.

How many lights were there above the sidewalk in the vehicular tunnel? One individual remembers there being one hundred and thirty-eight. Another person remembers there being one hundred and forty-seven.

The official opening of the tunnel was on September 30, 1939. However, it was actually opened for traffic almost eight months earlier and some kids may have walked through it even earlier. Garland Bray can remember going to a Frankenstein movie at the Princess Theater down in Bingham and then using a flashlight to walk back home through the tunnel before it had overhead lights installed. He was nine years old at the time and, after the Frankenstein movie, it was a bit scary to walk through the dark tunnel.

Figure 60 – Upper Portal Vehicular Tunnel 1947
Courtesy of Kennecott Utah Copper - The dome-shaped fan structure is to the right with the opening in shadow, the door to the control room is below the traffic lights, the roll-down articulated door is at the top of the entrance and the revolving door for pedestrians is to the left. A Grey Line tour bus is on the left. Boys would offer to guide tourists to the best points for observing the mine and would describe the mine in much the same detail as the Utah Copper representatives at the official Observation Point. These young guides would generally take their customers to unofficial observation points high up on the West Side of the mine from which the view was truly spectacular. Individuals in this photo include Eldon Bray in sport coat in front of the tunnel and Max Ivie immediately to his left. Other two boys are unidentified.

Figure 61 – Demolition of the E-Line Bridge About 1940

From the Gordon Bodily Collection - The E-Line Bridge was considered to be an engineering marvel. It was one of three bridges in Copperfield that provided access in the early days of the mine for trains from the West Side of the open pit to the levels on the East Side primarily for haulage of waste rock to the East Side dumps. The East Line Extension Bridge and the Upper Crossing (H-Level wooden bridge) had previously been demolished. The road from Bingham to Copperfield went under this bridge so it had to be maintained until after the Bingham-to-Copperfield vehicular tunnel was completed in 1939. This photo was taken in about 1940.

CHAPTER THREE

FEATURES OF COPPERFIELD

There were more than a dozen significant features or activities in Copperfield that made Copperfield what it was. We took most of them for granted. My list of features, not necessarily in order of importance, include the: 1) The Vehicular Tunnel (Bingham-Copperfield Tunnel), 2) Hills of Bingham, 3) Utah Copper open-pit mine, 4) Underground mines, 5) Neighborhoods, 6) Businesses, 7) Grade School, 8) U.S. Hotel, 9) Mine Observation Point, 10) School buses, 11) Bingham High School in Copperton, 12) People, 13) Sleigh riding and 14) Boy Scout Troop 112.

Most of these features were of different importance to each person depending on his age, personal interests or activities and time of year. One extreme might be a girl who was strictly supervised by her parents and almost never ventured from her home except to go to school. The other extreme might be a boy who spent most of his spare time, unsupervised, outside of his home. They would have vastly different experiences even though they lived in the same town at the same time.

People resided in Copperfield from about 1870 to 1958. My years in Copperfield from 1935 to 1952 were near the end of its 90-year history. The early heyday years of underground mining were nearing their end and much of that history had already been lost from the memories of most Copperfield residents. The U.S. Company's underground mine in Galena Gulch was being phased out, the Niagara Tunnel in Copperfield was still in operation but much of the U.S. Company's mining was being conducted from Lark through the Mascotte Tunnel. Dozens of other underground mines had been shut down decades earlier.

The "Shady Track" (which no longer had any railroad tracks and along which I walked many times) was actually the "Shay Track", part of the Copper Belt Railroad from 1905 on which trains of small ore cars pulled by Shay engines once ran. The "Cocoa Dirt" was actually the tailings from an old mill that used to be in Telegraph. The old "trail" (abandoned rail line) above the road across from the "Cocoa Dirt" was the Holden Tramway rail line. The U.S. area above Copperfield was actually the site of the first mining claim in Utah in Galena Gulch or Upper Bingham in 1863. The "Horseshoe Bend" where we used to hunt cottontails was the Horseshoe Bend on the High Line of the Rio Grande Western Railroad that hauled ore from Bingham but the tracks and trestles had long since been removed.

Many of the gulches were known by the name of the mine that was in operation there. The "U.S." was actually Galena Gulch, the "Queen" was Blackjack Gulch, the "Fortuna" was Keystone Gulch, "Big Tree" was in Bear Gulch and the "Silver Shield" was Porcupine Gulch. There were still some wooden signs with Chinese characters nailed onto trees along the old timber trail that ran from just above the spring at the Middle Canyon-Butterfield saddle to Galena Gulch. The trails were used to haul timber for use as supports in the underground mines in Galena Gulch and Upper Bingham.

THE VEHICULAR TUNNEL IN OUR LIVES

The Bingham-Copperfield vehicular tunnel, after it was opened to traffic on February 4, 1939 (it was dedicated more than seven months later on September 30, 1939), was an everyday necessity in our lives. It provided essentially the only road access to Copperfield. The old road between the main part of Bingham and Copperfield, including the E-Line bridge, was mined out shortly after completion of the tunnel. The mile and a quarter long tunnel was one-way only – this entered the additional factor into our lives of having to wait for the traffic light to turn green. Most Copperfield residents spent many hours total, in five to thirty minute periods, waiting at the lower or upper entrances to the tunnel for a green light. Hopefully, due to good luck or planning, no wait would be necessary. A round trip from Bingham up to Copperfield and back could be facilitated by a quick turnaround in Copperfield so as to catch the green light before it turned red again. Most of the time though it was necessary to wait for the green light.

A description of the tunnel, from the viewpoint of an everyday user is as follows. The tunnel was wide enough for a bus with about eighteen inches to spare on each side – a car had more room to spare but still caused anxiety for many tourists. The direction of traffic flow was controlled by a set of electric eyes near each entrance. There were three turnouts at roughly equal intervals; each turnout had room for three cars to squeeze in without blocking the roadway. A walkway with a concrete and iron pipe guard rail was elevated above the road level on one side of the tunnel. There were either one hundred thirty-eight or one hundred forty-seven incandescent lights (according to two people's memories more than fifty years later), such as in a house, attached to the ceiling immediately above the walkway. They were always lit. A stream of water (stormwater and sewer) ran under the walkway about four feet below the road. A huge electric exhaust fan was located at the upper end of the tunnel in Copperfield. There was a reticulated door at the top of the tunnel entrance that closed the entrance off when the fan was operating. A revolving door allowed pedestrians to exit the upper entrance when the main entrance was closed off.

If the exhaust gases in the tunnel became too thick then the reticulated door would roll downwards to close off the entrance and the fan would automatically be turned on and would suck the gases out. The fan would run for about twenty minutes and would do a reasonably good job of sucking out the exhaust fumes. Ideally there wouldn't be any cars in the tunnel when the fan came on. The traffic lights would stay red at both ends until all the cars were out and would remain red until the fan completed its operation. During the summer when there was heavy tourist traffic the air in the tunnel would become so full of automobile exhaust fumes that the overhead lights would look orange. The fan would come on before the gas was truly dangerous but we did walk through the tunnel several times when the lights were orange – I have no idea what the concentration of carbon monoxide was at that point.

Judging from the description of the tunnel one would imagine a smooth operation with waits of about twenty minutes if a car arrived at an entrance just after the traffic light turned red. It would take about ten minutes for all the cars to drive in one direction and another ten minutes for the cars that had been waiting at the opposite entrance to come through. During the day shift, especially during the summer when traffic was heavy, a watchman or traffic director was generally on duty. This traffic director would wave for the cars to stop or go and would sometimes manipulate the traffic lights.

There were, however, several problems that could foul up the whole operation. For example, if there was no traffic director on duty and a string of slow-moving cars had entered the tunnel just before the light turned red, then there would still be time for somebody to drive more rapidly and catch up with them before they exited at the other end. This would save twenty minutes of waiting time for the impatient driver. This practice could be extended so that if one could still see the taillights of the cars before they went around the curve in the tunnel then there was still time for a fast driver to catch up. One could extend it still more so that if one thought the cars had just barely gone around the curve then he could still catch up with them.

Sometimes the electric eyes or other mechanism would go awry so that the traffic light would remain red at both entrances for a prolonged period. If one "suspected" that the lights weren't operating properly then he might go into the tunnel even though the light was red. Sometimes two drivers or strings of cars at opposite ends of the tunnel would decide the lights weren't working and both would enter the tunnel only to meet each other part way through. Needless to say, it was very common for one string of cars to exit the tunnel and another to start in while there were still some latecomers who they would meet head to head in the tunnel.

When cars going in opposite directions met in the tunnel then they obviously had to stop and figure out their next move. Usually one group of cars would be much closer to a turnout or to an entrance than the other group. Most often the cars closest to the turnout would back up and pull into the turnout (which could hold three cars maximum). Then the other cars would drive past those in the turnout with no further difficulty. If there were more than three cars in both groups then it might require some see-sawing back and forth before all the cars were free to continue on to the exit.

Sometimes there were some horrific traffic jams with the honking of horns and arguments as to who should back up. You guys ran the light! You're closer to the turnout than we are! There are more cars going uphill so you guys back up! I never witnessed any physical violence between drivers but there were lots of times when it was threatening. Sometimes things were so botched up that a bunch of cars might have to wait as long as an hour before they could drive into the tunnel.

I only witnessed one accident in the tunnel and I was involved in it (or my Dad was). My Dad, with Mom and I in the car, was the last in a line of cars going down toward Bingham. The traffic lights hadn't been operating correctly and we met another line of cars that were coming uphill. We all stopped. But then, from behind us, we heard another car coming at high speed. Duggie Gonzales was apparently trying to catch up with us and, as he rounded the curve, he saw us all stopped in front of him. He slammed on his brakes but wasn't able to stop in time. His car smashed into the rear of our car and buckled the bumper and rear end pretty badly so that it was essentially a wreck not worth repairing. All of us in both cars received quite a jolt but nobody was hurt.

Besides riding the school buses every day when we went to school in Copperton, we kids went through the tunnel about once a week and also played around or sold ore near the upper portal. Our perspective and some of our experiences with the tunnel were quite different from that of the grownups. Very few adults walked through the tunnel but we kids did it quite often – and sometimes in unique ways. Some kids walked through the tunnel backwards just to prove that they could do it. Other times we would run through the tunnel without stopping. And – we even jumped down onto the roadway and walked there instead of on the sidewalk. If any cars came while we were on the roadway we would hide behind one of the concrete pillars that held up the sidewalk – there was a big water pipe above the sewer water that we could crouch on. There were rats that lived in the sewer and I always worried about them when I hid from the cars.

Besides the legitimate problems with the traffic signals there were times when kids would crouch down below the electric eyes at the upper end of the tunnel and wave their hands across the beam just to see what the result would be. We walked and rode through the tunnel so many times that we knew almost every detail of it – even how many light bulbs there were.

When cars exited the Copperfield end of the tunnel they were going uphill so they would slow down quite readily and there was a seven-foot high concrete wall directly in front of them so they always turned down or up the canyon or stopped before hitting the wall.

The Bingham end of the tunnel, however, was a different situation. Cars were going downhill as they exited the lower end so there was a tendency for them to be moving faster – especially in the middle of the night when there was no other traffic and when they might have had too much to drink. There was a Greek coffee shop in line with the straight part of the tunnel at the lower end. A car came speeding out of the tunnel one night and just kept going straight into the coffee shop. It smashed in the front of the shop but there were no injuries. And then a year or so later another car did the same thing with the same result. After the second smash-up they cemented half a dozen rails into the ground to form a four-foot high barrier. This obviously solved the problem for the owner of the coffee shop. I don't recall anybody running into the steel rails.

One experience with the tunnel that I especially remember was when my friend George and I were setting off firecrackers. We saw an advertisement in a comic book for a large assortment of firecrackers and we sent away for them. It was against the law for us to have firecrackers – especially as big as some of those we bought. A couple of weeks after ordering them we received a notice that there was a box for us down at the freight depot in Bingham. We went down to pick up the box and were somewhat afraid because a sticker on the outside plainly said the box contained fireworks. But nobody tried to stop us and we brought the box home. There must have been twenty-five pounds of fireworks in the box and we divided them up between us. Then we proceeded to play with them wherever we went – specifically down by the tunnel. Somebody apparently called the police and told them that kids were playing with illegal firecrackers. Anyhow, we decided to set a big bulldog salute off just inside the tunnel entrance as the cars were about to exit. We halfway hid by the revolving door and peeked out to watch the excitement. The timing was perfect – the big firecracker went off right next to the first car just as it was coming out. But then we saw that the car was that of the Bingham City police. The loud explosion surely must have startled the two policemen in the car but we didn't wait to find out. We ran up the hill above the tunnel and didn't stop until we were half a mile away. We never did hear anything from the police – most likely they never did find out who had done it. In any event, just by driving up to Copperfield they had stopped the misbehavior.

When I was fifteen years old and learning to drive a car my Dad had me drive up through the tunnel while he sat in the passenger's seat. It was hard for me to judge just where the front fenders were in relation to the concrete walls of the tunnel but I tried to stay right in the middle. But my Dad started hollering "Move over! Move over!" I thought I was already as close to the left side as I should get but I moved over more to the left like I thought he was telling me. Then Pop really got excited! He hollered "The other way! The other way!" I must have been only a couple inches from the wall on the left side when I finally moved over to the right. Needless to say, Bingham was a hard place for a kid to learn how to drive but I finally did.

THE HILLS OF BINGHAM

The hills around Bingham are part of the Oquirrh Mountains. Bingham Canyon is located on the eastern side of the range and drains to the northeast down into the Salt Lake Valley. The peaks and ridges are somewhat worn and rounded and most of the hillsides are only moderately steep. The highest peaks that were within a day's hike for most Binghamites are Clipper Peak (elevation about 9,150 feet) and Sunshine Peak (West Mountain, 8,970 feet) above Highland Boy and Butterfield Peaks at 9,330 feet. The elevation and climate of the Oquirrh Mountains are very favorable for the growth of animal feed and cover. Essentially the entire mountain range is good deer and elk range. Most of the commercial timber was cut down more than a hundred years ago.

Copperfield was near the southern, highest reaches of Bingham Canyon. As Copperfield kids we generally hiked and hunted within five miles of our homes. There were quite a few areas and things of interest out in the hills.

The mountain side and foothills from the top of the ridge above the main part of Bingham (The Big Rocks) down to Copperton and Lark was a prime cottontail hunting area. The brushy draws around Horseshoe Bend were especially good. In the wintertime hundreds of deer would concentrate in The Big Rocks area near Fortuna (Keystone Gulch) so when we were hunting cottontails in the wintertime we would often see large numbers of deer. Coyotes, bobcats and mountain lions were also there but they were very elusive and seldom seen.

The higher-elevation areas above the U.S. (Galena Gulch) and of Butterfield and Middle Canyon were very enticing during the summer. They were cooler and greener with some beautiful views and we could hike and hunt to our heart's content. We didn't shoot cottontails during the summer so we would hunt squirrels and rock chucks or just hike around.

Big Tree was a very large cottonwood tree situated in a large open flat in the bottom of Bear Gulch where several side gulches came in. It was a popular place because it wasn't too far from town but was far enough away to offer quiet and solitude and a nice walk. People would go there to have picnics. It must have been a popular spot for the Indians also because some arrowheads and other flint articles were found there.

Butterfield Canyon had a running stream of water that was an attraction so we would hike over there quite often. At one time several hundred rainbow trout were planted in the tiny creek but they didn't last very long. We would often gather watercress from the creek. We would sometimes camp out and play games in the large open area that we called "The Scout Camp". At the bottom of the clearing there was a large iron pipe from which cold, clear water poured forth.

There were a lot of chokecherry bushes near Copperfield and we would often eat so many of them that our teeth would turn black. There were also quite a few service berries bushes and one pinon pine tree. The pinon pine had good nuts and we buried some of them in the ground with the hope that more trees would sprout from them but I never saw whether this effort succeeded or not.

DEPUTIES WERE HERE ON A WILD GOOSE CHASE

Deputies Joe Raleigh and Dick Richardson of Salt Lake were up here on Wednesday evening and night on what might be called a wild goose chase. They came out here for George Lazos, violently demented man of Copperfield, but George was not there and the officers were informed that he was in Salt Lake. They came back to Bingham and called Salt Lake and asked about the matter. After some little investigation it was discovered that the man was in jail here, and was at the time the officers were dispatched to Bingham. There was an auto driver and two officers who made the trip out here, and it is a matter of fact that some expense will be attached to the affair. When this bill comes up for adjustment it will raise a nice point of law as to where it should be charged. Some one blundered, but the county will doubtless stand the expense without registering any kick.

The above article was printed in the April 12, 1918 edition of The Press-Bulletin, Bingham Canyon, Utah.

Figure 62 - Utah Copper Mine, Copperfield and the Hills About 1946
From the Gordon Bodily Collection (now the Dee Johanson Collection at the Utah State Historical Society) – Copperfield is on the far side of the open-pit mine near the center of the photo. The hills on the far side of the mine (south) and to the left of the mine (east) on the valley side of the mountain are where we spent many days hunting and hiking. Middle Canyon is to the right of the photo and Horseshoe Bend is to the left of the photo; we hiked and hunted in these areas also. We generally didn't go more than five miles away from Copperfield. The upper part of this photo shows parts of Utah Valley and Cedar Valley, and the Traverse Range which we did not walk to. The hillsides in the foreground and right are the upper reaches of Markham and Freeman Gulches that we kids from Copperfield seldom explored (see "A Bulldozer Load of Deer").

The East Side levels are to the top of the ridge (Q-Level) which indicates this photo was 1944 or later. The large, level area at the bottom of the pit was 6090 and the trench-like drop cut is for the 6040 tunnel that was completed in 1946. Mining of the bottom of the pit hasn't progressed beyond the drop-cut and trench to the tunnel; this photo was taken in about 1946.

Figure 63 - Utah Copper Mine, Copperfield and the Hills In 1957
Photo Courtesy of Kennecott Utah Copper – This photo was taken one year before the town of Copperfield was demolished. Compared to the 1946 photo, in 1957 the open-pit mine was much deeper, Greek Camp and all other downtown stores or structures such as the Pan-Hellenic Grocery and Joe Berger's Combination Bar were much nearer to the very edge of the pit. The hills, however, were much the same except that the mine had chewed a bit more of them away immediately adjacent to the pit and the waste dumps covered a larger area. The waste dumps in the foreground had dozens of impoundments (leach ponds) but only a few of them were filled with water at this time. Bear Gulch (Big Tree), Silver Shield and other parts of the hills that we used to roam were still untouched. Deepening of the pit to mine copper ore will require that more waste be mined from the upper levels so that the slope doesn't become too steep. The communities of Copperfield, Highland Boy, Carr Fork, and Bingham stand in the way of digging waste from the upper levels. The buildings in Copperfield will be demolished in 1958. Buildings in the other communities will also be demolished a few years later. The distance that waste rock has to be hauled has also increased over recent years and the main Bingham Canyon will be a handy place for waste rock to be dumped.

THE OPEN-PIT MINE

The town of Copperfield and the lives of its inhabitants were dominated by the open-pit mine. Most of the men were employed by the Utah Copper Company (later Kennecott) so that the economic welfare of almost the entire town was dependent on it. The operations of the minc were in evidence all around the town – parts of the mine could be seen from the front yard of most homes, the main road down the canyon was cut off by the huge open pit, trains ran along H-Level and through Fox's Switch. The Mine Observation Point was at the end of Main Street below the vehicular tunnel; the huge open pit was immediately below that. The afternoon dynamite blasts that reverberated through the canyon were hardly noticed.

Figure 64 – Electric Locomotive w Ore Cars Being Loaded By Electric Shovel in 1936
From the Gordon Bodily Collection – The electric locomotives and the ore cars shown here were very familiar to the residents of Copperfield. The electric shovel, with "caterpillar tracks" had replaced the earlier steam shovels that moved on railroad wheels. However, this Standard shovel, on which only the "circle" and the boom rotated, was replaced during the life of Copperfield by "Whirlies" on which the entire body of the shovel rotated. The major technological advance of the replacement of rail cars by large haulage trucks would begin in 1963.

Figure 65 – Copperfield on the Edge of the Utah Copper Open-Pit Mine About 1941
From the Gordon Bodily Collection - This photo is looking southeast from Carr Fork across the mine to Copperfield. The roadway from Bingham to Copperfield and the E-line Bridge have been mined out. Some of the buildings on the Main Street of Copperfield are near the edge of the pit. The houses of Dinkeyville and Terrace Heights are farther up the canyon. H-Level stands out as the prominent level above the Terrace Heights and below Dinkeyville. The mine levels on the East Side have been extended up to L-Level. The grade is being prepared for N-Level. O, P, and Q-Levels have not been started; the top of the ridge is still virgin country. The bottom of the pit is 6190 or 6140. Mount Timpanogos and other peaks on the east side of the Utah Valley are in the background. This photo was taken in about 1941.

The Utah Copper Company affected the lives of the people in Copperfield and the whole of Bingham in many different ways. In general the company was a very good neighbor that helped the people to prosper. Most of our Dads worked for the Utah Copper and so we depended on it for food on the table and for roofs over our heads. Indeed, most of the houses in Copperfield were owned by the Utah Copper Company from whom they were rented by our Dads.

We kids were also affected in other ways by the presence of the mine. Legitimate ways in which the mine affected our lives were that we sold ore to the tourists who came to see the mine, for recreation we went to the Gemmell Club that was owned and managed by the company, the Utah Copper supported some activities of the Boy Scout Troop (especially by continuing to pay the regular salary of the Scoutmaster when we went to Camp Steiner) and almost every day there was the sound of blasting and the movement of trains along H-Level and through Fox's Switch.

Figure 66 – The Utah Copper Pit in 1946 With Bingham and Copperfield On Periphery
From the Gordon Bodily Collection - Bingham and Copperfield are on opposite edges of the pit; no vestige remains of the roadway between the two. The 6040 tunnel has been completed and is in use. The bottom of the pit is essentially at the 6040 elevation all the way across. All the levels on the East Side (Q being the highest) have been established. This is the mine as most of us who grew up in Copperfield remember it – it continued to grow deeper and wider and to chew away more of the town year by year. Not only did it mine away the town but it also covered more and more of the hillsides farther down the canyon and around to the slopes of the valley with waste dumps. Waste dumps can be seen in the area of Dry Fork on the West Side of the mine. The Salt Lake Valley and the Wasatch Mountains are visible in background.

Illegitimate ways that the mine affected us were the attractions of property owned by the Utah Copper (and also by the U.S. company) that surrounded us on every side. We trespassed across company property when we took hikes or went hunting in the hills. We played on the coaches when they were tied up by Dinkeyville. Pennies and nails were set on the tracks so that the trains would flatten them out. We collected some ore samples from the open-pit mine although most of them came from the ore cars of the U.S. underground mine. We put iron nails into the "copper water" at the tailtrack of Fox's Switch so that the nails became coated with copper that we exhibited to tourists. We regularly swiped carbide from the underground mines. We even swiped dynamite and blasting caps from the open-pit mine but only a few times and the company would have immediately put a stop to this if they had known. The watchmen for Utah Copper were constantly chasing us off of their property. We actually did very little damage to company property but today, with lawsuits being so prevalent, our activities would have been looked upon more seriously.

UNDERGROUND MINING

The first mining claim in Utah, the West Jordan, was staked in 1863 near the head of Galena Gulch above what was Copperfield. Other claims soon followed and underground mines were developed. However, costs of transportation and supplies were so high that it wasn't until after the railroad was built in 1870 that underground mining flourished. Underground mining, with minor placer mining, was almost the exclusive method employed in all of Bingham Canyon from 1870 until 1906 when the Boston Con and Utah Copper open-cut mines were started. Dozens of underground mines operated in the Upper Bingham area with varying degrees of success from the 1860's to the late 1890's. In 1899 several of the larger, more important mines, including the Old Jordan and Galena, Spanish, and Telegraph mines were consolidated and the United States Mining Company was organized.

By the time I was growing up the underground mines in Galena Gulch (the U.S.) were being phased out with only a small amount of ore being mined by leasers. The only underground mining of real consequence in Copperfield was that of the U.S. Company through the Niagara Tunnel, the entrance of which was right behind the U.S. Hotel. Most of the men in Copperfield worked for the Utah Copper Company but about thirty who lived in Telegraph and Dinkeyville worked for the U.S.; in addition, the U.S. Hotel provided room and board for about fifty of their workers.

There were dozens or hundreds of abandoned mines and mine dumps in the Copperfield area. At least one man died by going into an adit that had bad air and at least one horse fell down a shaft and was killed. A person had to be careful around these abandoned workings and dumps. In spite of the hazards they were still generally of interest and curiosity.

The U.S. Company was also hauling considerable ore out through the Mascotte Tunnel to Lark but this had little impact on Copperfield. There were hundreds of miles of underground workings in the Bingham area. The U.S. Company's workings were interconnected by tunnels and shafts so that a person could travel from Galena Gulch to Copperfield or to Lark via underground workings in less than an hour.

A very important aspect of the U.S. Company's operation to the kids of Copperfield was that it provided most of the ore samples that they sold to tourists. Ore was hauled out through the Niagara Tunnel in small ore cars that were dumped at the trestle of the ore bins. The ore fell down into larger, standard-size rail cars that were underneath the trestle. The large ore cars were then hauled periodically to the mill and smelter down in the valley. The large ore cars were where the kids picked out the specimens which they sold.

An item of interest that a former worker over in Lark told me was the story of the miner with a wooden leg. While the men were gathered at the Mascotte Tunnel entrance before going in to work, this guy would clown around. His favorite trick was to stand on his wooden leg and grab the overhead trolley wire. His wooden leg provided insulation from the electricity. He was the only one who could get away with this. He would hold on to the live wire and grin at everybody. All was well until one day he forgot and let his other leg touch the ground. He was instantly knocked to the ground but escaped serious injury.

NEIGHBORHOODS OF COPPERFIELD

There were seven areas of Copperfield that were referred to by individual names. They were Main Street, Terrace Heights, Dinkeyville, Jap Camp, Greek Camp, Telegraph and the U.S.. The U.S. was torn down in about 1943; it was three miles or more above the main part of Copperfield. The great majority of the houses in Copperfield were owned by the Utah Copper Company except for Telegraph and the U.S. that were owned by the U.S. Company. The different parts of town were divided to some extent by races. Jap Camp housed most of the Japanese, Greek Camp was composed mostly of adult male Greeks (and some Puerto Ricans during and after the war) and most of the people of Mexican origin lived in Dinkeyville. The people of Main Street were mostly white but with some Greek families. Terrace Heights was primarily white but with some Mexicans. Telegraph and the U.S. were essentially all white. Each section of town played a different role in our lives depending upon our interests and needs. Where our friends lived was the major factor that determined how often we went to a particular part of town. No kids lived in Greek Camp and only a few lived in Jap Camp and Telegraph. About half a dozen kids lived in the U.S. before it was torn down. Main Street, Terrace Heights and Dinkeyville had the most kids.

COPPERFIELD IS VISITED BY DISASTROUS FIRE

Splendid School Building And Five Residences And Apartment House Destroyed By Midnight Conflagration – Tenants Were Aroused From Their Slumbers By Angry Flames – Loss Nearly $50,000 – Many Are Left Destitute.

The greatest fire in the history of Copperfield and the most disastrous in the entire camp since 1895 broke out in an apartment house shortly after midnight Wednesday morning, destroying two large apartment houses, three residences, the splendid school building and partially destroyed two other residences, entailing a loss approximately $50,000.

The fire originated in the rear of one of the Caulfield apartment houses presumably by a spark from a passing locomotive, and had made great headway before it was discovered. There were eight families living in the apartment houses and many of the occupants had barely time to escape into the street, leaving their belongings to be consumed by the flames. Between 60 and 100 were made homeless and many of those were left without clothes.

The flames spread rapidly over the dry wooden structures and but for the heroic work of the firemen may have swept the entire town. The Copperfield firemen commenced their work at once and a call was sent to Lower Bingham and the two fire departments from here responded at once. It was a desperate fight combatting the flames reaching into the heavens to save adjoining buildings. Before the fury of the conflagration was checked the flames leaped across the street and caught the residences adjoining the school building and then the school building itself caught fire.

Owing to the location of the school building the water pressure was not sufficient to throw an effective stream on the roof and could not be saved. But the water was used with telling effect on the buildings on either side of the apartment houses. Such floods were turned on these buildings that in spite of the intense heat no other structures on that side of the street were totally destroyed and this probably saved the town from being swept off the map. The Bingham firemen were not able to get in their best work when they arrived owing to the fact that the Copperfield and Bingham connections are not uniform. It is understood that they have reducers to use in case of such emergency, but they could not be located at the time.

The losses consisted of the school building which cost about $20,000 but which could not be replaced for less than $25,000 or $30,000; two residences belonging to J. Richards, one residence belonging to George Simmons, and two residences belonging to Mr. Caulfield partly destroyed. No estimate has been placed on the furniture and other belongings of the tenants, but it would reach well into the thousands.

A committee was appointed Wednesday to solicit contributions for those made destitute by the fire and the people in the camp are responding freely to their cause.

The appearance of the fire stricken district next morning was indeed ghastly. The bleak walls of the school building were still standing, the debris of the apartment houses were still standing, many firemen who had fought the flames and rescued children during the night were still on the scene viewing the havoc.

The school building was insured under a blanket policy which is understood to be about the value of the structure at the time it was erected. The Caulfield property was insured for $6,000. The other property was without insurance.

The school building will be rebuilt immediately according to information received from F. W. Quinn, member of the district board of education, and it will be modern in every respect with steam heat and other conveniences. Some were inclined to believe that the walls of the old building were uninjured and might be built to but it is understood that the board of education will not take any chances and will erect a new building outright. Every possible effort will be made to see that the school children of Copperfield will be cared for when the next session of school convenes.

The above article was printed in the July 5, 1918 edition of The Press-Bulletin, Bingham Canyon, Utah.

Figure 67 – 1st of a 4-Photo Panorama of Copperfield and the Mine in 1926
Used by Permission, Utah State Historical Society, all rights reserved, photo is from the Shipler Collection – Dinkeyville is in the left and foreground of this photo; a steam engine with five loaded cars is on H-Level above Terrace Heights. The ore bins that emptied down to H-Level and the abandoned Holden Tramway are at the far end of Dinkeyville. Telegraph is in the upper center of the photo. The Upper Terrace Heights two-story duplexes and both Terraces of one-story duplexes can be seen. The schoolhouse and some of the houses along Main Street are shown. The addition to the rear of the schoolhouse is under construction. In the tailtrack of Fox's Switch, by the Copper Water, there is a steam engine with eleven loaded ore cars. Four wagons pulled by horses are on the main road just above the Copper Water. Above Fox's Switch, just below the Jap Camp Road, is a mine with an ore bin and two ore chutes that empty down to Fox's Switch. The compressed air line from the U.S. Niagara Tunnel goes through the concrete H-Level underpass, up the road and then up the hill to the U.S. mines in Galena Gulch. Many of these interesting features are only evident when the photo is enlarged. A portion of this photo was used in the front cover

Figure 67 (continued) 2nd of a 4-Photo Panorama – Part of the Upper Terrace and the west side of Upper Main Street from Fox's Switch to the Circle and to the U.S. Hotel are in the foreground. Jap Camp does not have the shower house yet; in its place is a two-man outhouse. The U.S. Hotel is in the front center of the photo with the ore bins to the right. There is no ore chute to the right of the hotel. The U.S. Company's heating plant and air compressor are immediately behind the hotel. The compressor primarily provides air to the company's underground workings through the Niagara Tunnel but an above-ground line from the compressor also supplies air to the company's mines in Galena Gulch. The above-ground compressed air line can be seen where it crosses Jap Camp on wooden supports by Miya's boarding house. The portal of the Niagara Tunnel is beneath the large dry/change house with stacks of timbers to the right. Copper Center Gulch, which is filled in with waste rock, is immediately behind the Niagara Tunnel entrance.

Figure 67 (continued) 3rd of a 4-Photo Panorama – The ore bins of the Niagara Tunnel are in the lower left of the photo with Greek Camp immediately above. The southwest side of the open-pit mining operation is shown. The mine operations are a considerable distance away from Greek Camp – the pit isn't threatening to dig it away yet. There are about a dozen shovels in the photo. The shovels are all electric – they don't have smokestacks, are not emitting any smoke, and there are towers on the levels that carry the electrical wiring. The shovels were recently converted to electrical power – in 1925 and earlier in 1926. The engines, however, are still powered by coal – they are all giving off steam or smoke. Water tanks can be seen on the south end of most levels and several piles of coal are evident for servicing the steam engines. The steam engines will almost all be replaced by electrically powered locomotives by the year following this photo - in 1927.

Figure 67 (continued) 4th of a 4-Photo Panorama – The bottom of the open-pit is being pushed to the west into the higher grade center of the orebody. Whether ore or waste is being mined from a particular shovel location can be determined by the type of cars that are being loaded. The two shovels in the upper left of the photo are loading ore cars which are higher and not quite so long as waste cars. Also, the ore cars are constructed so that they are unloaded at the concentrator by being overturned whereas the waste cars are built so that they can be dumped to either side by compressed air cylinders while the wheels remain on the track. Two trains of waste cars are in the center of the photo near the large churn drill rig. The promontory on the lower left of the photo where the tracks have been cut off was probably the E-Level approach to the abandoned East Line Extension Bridge. An edge of the approach to the E-Line Bridge is on the lower left of the photo just above the muck pile. The waste dumps have not yet encroached very far on the mountainsides in the background. Markham Peak is near the top center of the photo with Freeman Peak to the right.

Figure 68 – Main Street, Terrace Heights and Dinkeyville Between 1928 and 1944
From the Gordon Bodily Collection - The U.S. Hotel and the dump from the Niagara (U.S.) Tunnel are in right foreground. The rows of company houses on both sides of Main Street can be seen. The Grade School (with the auditorium having been added) is in the center. Both the Upper and Lower Terrace Heights two-story duplexes are in the upper left. Part of Dinkeyville is also in the upper left-center. The old houses on the hillside below Terrace Heights have been removed – this indicates the photo was taken after 1927. The abandoned Shay Track of the Copper Belt Railroad is on the left of the photo just above the houses of Main Street and below Terrace Heights. In the far upper right the grade of the old Holden Tramway can be seen (we called it the "Dinkeyville to Telegraph trail"). The "shortcut" tunnel through the mountain to the Yosemite mine (Bishop's) was just above the ore bins by the Holden Tramway. The old ore bins, with their chutes on H-Level, and the trestle from the Dinkeyville road to the top of the bins are complete in this photo; no demolition of the trestle has taken place – the photo was therefore taken before 1944.

Figure 69 – Copperfield from Niagara Tunnel Area About 1953
Photo Courtesy of Shirl and Isabell Scroggin - The four big smokestacks in the center were for the coal-fired furnaces that powered the compressors of the U.S. (Niagara) mine and also provided heat for the U.S. Hotel and nearby houses that belonged to the U.S. Company. The Grade School and the roofs of houses along Main Street are in the foreground. The Terrace Heights, including the two rows of 2-story duplexes on the left and the two rows of 1-story duplexes behind the schoolhouse can be seen. Part of Dinkeyville is also shown. The abandoned Shay Track of the Copper Belt Railroad is on the left between Main Street and Terrace Heights. The railroad in the right foreground goes up to Fox's Switch. The new roadway from just above Dinkeyville to the Copperfield-Lark saddle indicates this photo was probably taken in about 1953.

Figure 70 – Main St., U.S. Hotel to Telegraph w Jap Camp; View Up Canyon After 1945
Photo Courtesy of Kennecott Utah Copper - The portal of the Niagara Tunnel of the U.S. Mine is beneath the large building (Change House) near the right side of and a third of the way up the photo. Jap Camp is near the center of the photo; Fox's Switch is immediately below Jap Camp. The U.S. Hotel is in the lower right. Houses of Main Street are to the left (up the street) of the hotel. The backs of the two-story duplex houses of Terrace Heights are in the center foreground. The early-day waste dumps from the open-cut mine extend from the Niagara Tunnel all the way to the skyline (where Copper Center Gulch used to be). These old waste dumps are being mined away in the upper right of the photo. The waste dumps also extend to the left around the mountain into Galena Gulch. Leaching of copper from these old dumps was responsible for the "Copper Water" which is at the tailtrack of Fox's Switch on the left of the photo. The large metal building on the West Side above the center of the photo is on about J-Level. It was used as a materials shed. A similar building on the East Side on the I-J tailtrack is near the bottom left of this photo; it was not present in 1944. The ore bins that used to be on H-Level near the left side of the photo have been removed.

Figure 71 – Upper Copperfield, East Side of Mine to N-Level In 1934
From the Gordon Bodily Collection - Dinkeyville, Terrace Heights and part of upper Main Street (including the schoolhouse) are visible. The abandoned bed of the Holden Tramway is on the right side of the photo above the roadways; the tramway can be traced around the hill into Dinkeyville with a wide U-curve between H-Level (the major level in the bottom of the photo) and the I-Level tailtrack . The roadbed of the abandoned Shay Track is below the Lower Terrace Heights. The L-N switchback has been completed and N-Level is the highest level on the East Side of the mine. The "Upper Crossing" wooden bridge on H-Level was replaced in about 1925 by the concrete overpass. The huge waste dump from the early Utah Copper that filled the north side and bottom of Galena Gulch is on the left of the photo. This dump covered several historical features such as the D&RGW High Line railroad that went up to the U.S. Mine (Jordan and Galena) and part of the aerial tramway that came down from the U.S. Mine through Copperfield to the railroad in Bingham. The ore bins on H-Level halfway up the right side of the photo are still in place.

Figure 72 – Upper Copperfield, East Side of Mine to Q-Level (Top of Ridge) In 1944
From the Gordon Bodily Collection - Dinkeyville, Terrace Heights and upper Main Street are visible. The "Dinkeyville to Telegraph" trail (bed of the Holden Tramway) is in center-right of photo above H-Level which crosses the canyon via the concrete overpass. The "Shady Track" (road bed of the Shay Track) runs below the Lower Terrace Heights and above the houses and school on Main Street. This photo was taken ten years after the previous one from almost the same location. The biggest change from 1934 is the development of mine levels all the way to the top of the ridge in this photo. The trestle of the ore bins (between H-Level and the Dinkeyville road) is partially dismantled but the ore bins themselves are still in place. There is no large metal building on the East Side I-J tailtrack – its presence can be used to date photos.

Figure 73 – Dinkeyville, Terrace Heights, East Hillside of Copperfield In 1953
Photo Courtesy of Kennecott Utah Copper – Dinkeyville is in the center of the photo with H-Level separating it from the Terrace Heights which is immediately below. The schoolhouse and the houses of upper Main Street are along the left half of the bottom of the photo. The Shay Track is above the houses along Main Street; the roadway to the Lower Terrace, behind the schoolhouse, essentially follows the route of the old Shay Track. The large metal building in the foreground is on J-Level. A similar building is on the East Side on I-J tailtrack (the first one above H-Level) at the same elevation as Gus the Goatman's place. The "Saddle" or divide between Copperfield and Copper Gulch is on the skyline. The switchbacks and levels of the mine are to the left of the Saddle and the hillside to the right is relatively untouched by the mining operation. A new roadway has been constructed above Dinkeyville to the J-K switchback and then up to the saddle. The ore bins that used to empty onto H-Level on the right side of the photo have been completely removed.

MAIN STREET

The Main Street section of Copperfield ran along the bottom of the canyon from the edge of the open-pit mine up to the last residence by Fox's Switch. The lower one-third of Main Street was occupied mostly by businesses and the upper two-thirds by residences. There were about forty houses on Main Street. About a third of all the kids in Copperfield lived on the upper part of Main Street. Burkes, Tysons, Grahams, Katis, Robertsons, Barrets, Gardikas, Golishes, Jaterkas and Pantalones were among those who were close to my age. All of us walked down Main Street or played there fairly often. The business district, observation point, vehicular tunnel, U.S. Hotel, and the Copperfield Elementary School were all on Main Street. We caught the school bus by the portal to the tunnel. We often played games like kick-the-can or touch football on the Main Street or in the alley by the U.S. Hotel. The school ground was also used for games quite often. After dark, for entertainment, we sometimes knocked on doors or swiped apples from the two houses that had apple trees. In the wintertime we spent a lot of time riding sleighs down Main Street.

Figure 74 – Fox's Switch, Fox's Boarding House, Main Road, Grade School About 1919
Photo From Bingham High School Alumni Collection – This "photo" is quite unique in several respects. It shows Fox's Switch and Mrs. Fox's Boarding House (in the foreground) after which the

switch was named. Between the railroad and the boarding house is a long sluice-like structure that might have been used to precipitate copper or to recover placer gold. The roadway is unpaved. The school building is the new one that was built after the first one burned down in July of 1918; the chimney of the heating plant can be seen behind the rooftop of the school. The front steps of the school let down directly to the street. The rear addition to the school has not yet been built. Two large apartment houses across from the school and three houses adjacent to the school were burned down along with the original school. A surveyor with his transit is just to the left of the boarding house; he is probably surveying in preparation for road and building construction. The location of the original of this photo is unknown; the author would love to find a good copy of it.

Figure 75 – Upper Main Street With Fox's Boarding House and Fox's Switch About 1919
Photo Courtesy of Willimay Tervort, Shirl and Isabell Scroggin and of Max and Barbara Ivie - There are two vacant lots immediately below the boarding house as a result of the July 1918 fire; the previous figure shows them more clearly. The schoolhouse is just off the right side of the photo; the only visible sign of it in this photo is the flagpole. The U.S. Hotel has not yet been built. No Utah Copper houses are on Main Street. A building boom will commence very soon after this photo was taken. A horse and an early model truck are on the road just above the boarding house. Some of the houses of Jap Camp are in the upper left corner of the photo. The E-Line Bridge can be seen far down the canyon.

Figure 76 – Upper Main Street, Copperfield School, Copper Water and Overpass
Photo Courtesy of Shirl and Isabell Scroggin - The upper part of Main Street is lined with company houses on both sides; the schoolhouse is in left-center. Ted Scroggin's house is in lower left corner. The tailtrack of Fox's Switch is in the far right above center. H-Level is in the upper part of photo with the concrete overpass where it crosses the main road (see "No Brakes, Speeding Backward"). The "Copper Water" issues from the cleft part way up the embankment at the end of the tailtrack of Fox's Switch just above the houses on the right.

Tony Jaterka's house is the uppermost on the right just below the garages. Tony's apple tree was right next to his front porch and grew up past the roof. You could climb from the railroad embankment of Fox's Switch onto the back of his house, creep across the roof to the tree and then pick the apples. But Tony had keen hearing and if he heard anybody on his roof he would try to catch them. He would vault off the side of his porch and start running toward the back of the house. One of the kids would sound the alarm and we would run as fast as we could across his roof and jump over to the embankment and keep going from there. If he could hear us as we carefully crept along his roof then it must have sounded like a stampede as we ran back along the roof to escape from him. Tony never caught anybody that I know of.

The "Copper Water" was copper-bearing water that ran out of a small, old tunnel halfway up the steep wall at the end of the tailtrack of Fox's Switch. The old tunnel ran along the bottom of the old creek bed and captured the copper-laden water that drained through the old waste dumps above Jap Camp and in Galena Gulch. The water was deep green in color and was said to be poisonous. We would put iron nails into the water and in a few seconds they would be coated with copper.

Figure 77 – Circle on Main Street Copperfield Before Fence About 1922
Photo Courtesy of Shirl and Isabell Scroggin - The Circle extended back from the main street at a wide spot in the canyon (part of the mouth of Copper Center Gulch). This picture was taken at an early date when the fence around the "Circle" had not yet been built. The right side of the duplex immediately behind the Circle, facing the street, was rented by Liston Bray when he went to work for the Utah Copper at age 16 in 1922. His mother's family lived there and also took in boarders. When Liston retired, at the age of 65, he had worked for the Utah Copper and Kennecott for 49 years.

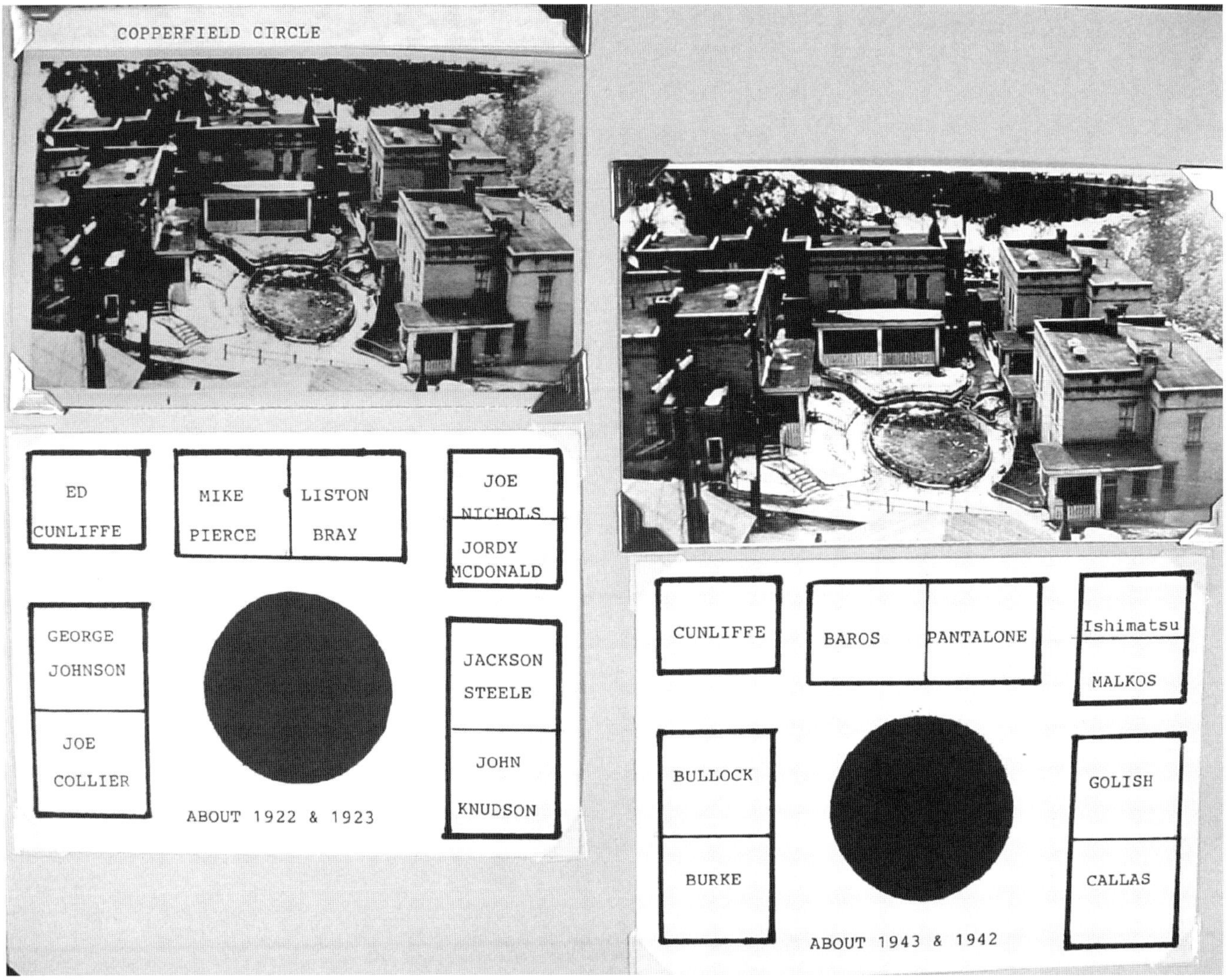

Figure 78 – Occupants of The Circle About 1922-1923 and 1942-1943
Photo Courtesy of Shirl and Isabell Scroggin – The nine homes or apartments in the Circle were all owned by the Utah Copper Company (later Kennecott). They were built in 1922 and rented to various employees as vacancies occurred. Of the nine residents in 1922-23 only Ed Cunliffe still lived there in 1942-43. Over the thirty-six year life of The Circle dozens of different families called it their home. Arilla Bullock Jackson's (author of "Copperfield" 1984) home from 1929 to 1937 was on the left (up-canyon) side of The Circle. Maintenance of the houses was performed by the Utah Copper Company. A crew of maintenance men cleaned the wallpaper, painted the houses and did other necessary work on a periodic basis.

Figure 79 – Uppermost Main St. with Fox's Switch and Terrace Heights About 1941
Photo Courtesy Of Kennecott Utah Copper - At far left is Fox's Switch with ore cars for Leasers (Independent miners who leased the rights to mine ore in certain areas from the U.S. Company). Many of the houses on both sides of Main Street can be seen as well as the top of the U.S. Hotel. The schoolhouse is on the right (east) side of the canyon. Terrace Heights is on the right side of the canyon above the schoolhouse. The first house on the left (west) side of Main Street, below the garages, with the apple tree growing in the front yard was Tony Jaterka's in the 1940s. Apples could be reached from the roof of his house (if you were faster than Tony). The model years of the three automobiles in the foreground help to date this photo.

TERRACE HEIGHTS

The Terrace Heights was on the east side of the canyon above Main Street but below Dinkeyville. It included four rows of duplex houses. Two of the rows each had six two-story duplexes; the other two rows had one-story duplexes – the top row two duplexes, the bottom row three. The total number of family units on Terrace Heights was thirty-four. My Dad rented half of a two-story duplex on the Upper Terrace (691-A) for $20 a month which they took out of his paycheck. The Alton Ivie family lived in the other half of the duplex (691-B).

The Terrace Heights had no community area to serve as a playground so we played in our small yards or went elsewhere to play (such as to the school grounds). One advantage of the Upper Terrace was that it afforded ready access to the hills up behind Copperfield. There wasn't a road behind the Upper Terrace so coal was delivered to the bottom of the incline on the road below and then was carried up the incline and to the individual homes by wheelbarrow or by hand. Homes on the Lower Terrace had a road behind them so that coal could be unloaded right next to the coal chute. From there the coal only had to be tossed into the coal chute and it would slide down to the coal bin right next to the house.

Kids who lived on the Terrace Heights that were near my age included those in the Ivie, Barton, Gonzales, Miller and Palmer families. There was a shifting of families within the town (generally out of the houses higher up the hillside and into the houses on Main Street). Quite a few families moved clear out of Copperfield when I was growing up there; most of them moved to Copperton or all the way down to communities in the valley such as Midvale, West Jordan, or Sandy.

There were eight garages for the Terrace Heights. They were located on the road below the Upper Terrace halfway between the one-story duplexes and the two-story duplexes. These garages led to at least one humorous incident. The garages were quite narrow so that one had to be careful not to hit the wall on either side when driving in or out of the garage. The roadway in front of the garages was also quite narrow and dropped immediately off onto the steep hillside below. The hillside ran all the way down to the schoolyard in the bottom of the canyon with only the narrow sidewalk for the Lower Terrace and a couple other trails as minor breaks in this steep dropoff. Along the edge of the roadway there was a barrier of vertical iron rails about three feet high placed fifteen feet apart with two steel cables running through them. This barrier was intended to prevent automobiles from driving off the road and falling down the hill.

When coming out of the garage you couldn't just head out, take a wide swing and drive down the road. you had to slowly drive forward to the barrier, stop, back up while turning the wheel clockwise, stop again and then make a left turn as you drove forward and along the roadway. This sequence of careful maneuvering proved to be too much for one lady. She drive straight out of the garage, hit the barrier and the front end of the car started down the hill. Luckily the series of iron rails and the steel cables were strong enough that they stopped the car and held it so that the front end only went down the hill about six feet. The lady was thoroughly scared and a crowd of bystanders quickly gathered to help, offer advice and to watch. A wrecker had to be used to get the car back up and onto the roadway. Whether the lady continued to drive after that mishap isn't known but nobody could blame her if she quit.

Figure 80 – The Grade School, Upper Main St, Terrace Heights and Dinkeyville 9-14-25
From the Gordon Bodily Collection – The Grade School does not have the extension built onto the back yet – it will be built in 1926. The single family company houses along Main Street were completed a year or more earlier. The one-story duplexes of Terrace Heights in the upper right of the photo were built in 1924. The two-story duplexes of the Upper Terrace Heights are under construction. Work has not yet been started on the two-story duplexes of the Lower Terrace Heights. Some of the wooden houses in Dinkeyville can be seen. One structure of special interest is a tower or station for the aerial tramway that carried ore from Galena Gulch (The U.S.) to Lower Bingham – it is near the upper right corner of the photo just above the one-story duplexes. Another station (evident in a 1908 photo) was a few yards east of the main road up the canyon about eighty yards below the Upper Crossing (H-Level), just below the Shay Track (which became the road to the parking area behind the Grade School); both of these stations would be demolished before 1934. Four aerial tramway stations can be seen in the 1909 photo of the Old Jordan and Galena Mines. These few stations only generally mark the route of the aerial tramway that traversed Copperfield.

Figure 81 – The Grade School, Upper Main St, Terrace Heights and Dinkeyville 9-14-25
From the Gordon Bodily Collection – Construction of the Upper Terrace two-story duplexes was started in 1925 and they are essentially completed in this photo. Excavation and site preparation for the Lower Terrace has been started. The older wooden houses have been removed from the immediate area of the Lower Terrace but several of them remain along the bottom of the photo – they will be demolished within a year. H-Level is immediately behind the Upper Terrace houses; it separated the Terrace Heights from Dinkeyville. Half a dozen of the wooden homes of Dinkeyville are shown in the photo; some will be demolished within a year or two while others will endure until about 1957.

Figure 82 – Copperfield About 1927, Lower Terrace Heights Being Built
From Grace Bray – The two rows of one-story duplexes, to the right above the schoolhouse, were built first, about 1924. This photo shows the Upper Terrace of six two-story duplexes was already completed while the Lower Terrace was still under construction; the farthest right of the bottom row of six duplexes may not been started yet. The row of garages between the Upper and Lower Terraces hasn't been built yet. The older wooden houses immediately below the Lower Terrace haven't been removed. The old Shay Track is just above the houses on Main Street and below the Lower Terrace. The addition to the schoolhouse has been built but the retaining wall and parking lot behind the addition have not. The top of the U.S. ore bins, where ore from the Niagara Tunnel was unloaded, is in the left foreground with the roof of the U.S. Hotel behind it.

Figure 83 – Terrace Heights With Parts of Dinkeyville and Main Street in 1927
From the Gordon Bodily Collection - These two terraces, each of six 2-story duplexes housed twenty-four families. There were an additional two terraces of 1-story duplexes to the right of this photo. Some old houses were demolished to make room for Terrace Heights; some of the old houses still remain below the Terraces on the left side of the photo. The Shay Track (Copper Belt Railroad) bed can be seen above the houses of Main Street, below the Terrace Heights; we called it "The Shady Track". H-Level of the open-pit mine is immediately above the Upper Terrace. Three tailtracks of the upper levels of the open pit on the East Side are above H-Level; they would be I-J, J-K, and K-L tailtracks. The upper part of the ridge is still relatively untrammeled. Tommy's Rock was above the Terrace Heights and Dinkeyville halfway between the uppermost tailtrack and the top of the ridge. The house on the Upper Terrace, fourth from the left, third from the right, was occupied by the LaMar and Grace Bray family (left side, 691-A) and the Alton and Allene Ivie family (right side, 691-B).

DINKEYVILLE

Dinkeyville was the farthest-up group of homes on the east side of the canyon above the Terrace Heights. There were about 30 houses in half a dozen rough rows or scattered groups. Most of the houses were one-story wooden structures. The paths and roadways were dirt or gravel. Many of the residents went to great effort to keep their homes as neat and comfortable as possible. About half the Dads in Dinkeyville worked for the Utah Copper and half for the U.S.. Some of the outstanding kids of about my age who lived in Dinkeyville were the Aguayo, Contreras, Cordova, Kubota, Leyba, Neria, Sanchez, and Valdez families.

Figure 84 – Dinkeyville and Most of Terrace Heights After 1927
Photo Courtesy Of Kennecott Utah Copper - Dinkeyville was so named because the small steam engines were called "Dinkeys" and there was a yard for them at the bottom of what was called "Dinkeyville", near where H-Level was in this photo. The two rows of one-story duplexes that were part of Terrace Heights are to the right of and on the same elevation as the two-story duplexes. The row of eight garages has been built on the road between the Upper and Lower Terraces. The "Saddle" or pass between Copperfield and Lark is at the top left. Torres's pig pens and field are halfway between the Saddle and Dinkeyville. Gus the Goatman's place is in the extreme left of the photo across from Torres's "farm". The overgrown trail that starts in the saddle and slants down to the right above Dinkeyville until it intersects H-Level on the right edge of the photo was probably part of the roadway or trail that went around the mountain to the Yosemite and Brooklyn mines.

An interesting resident of Dinkeyville, at least in my eyes, was "Gus the Goat Man". Gus had a herd of goats and lived in a ramshackle collection of tin-roofed goat pens and shacks on the outskirts of Dinkeyville. He made cheese from the goat milk and must have sold it and probably some young goats (mostly to other Greeks) to make a living. The goats were apparently his main source of income. He had four or five dogs that were trained to take some of the goats out each morning and bring them back each evening. Many of his goats, however, appeared to essentially run wild because a dozen or so could often be seen out in the hills by themselves several miles from Gus's place. The dogs might have been able to locate them miles away and herd them back home but I never witnessed this. He probably kept his milking goats in the pens and close to home while letting the nonmilkers use the hills farther away for their pasture. I doubt if these "wild" goats were able to fend for themselves indefinitely. I only saw them in the summertime and expect that the coyotes, lions and bobcats killed many of them while they were unattended in the hills. Gus and his dogs might have rounded up the survivors before the onset of winter.

My brother Garland recalls Gus inviting him and another boy, as they walked past his place, to come in. He gave them goat's cheese to eat and impressed them as a good man who was leading a life separate from and quite different than others. He must have been quite lonely. One picture I still have in my memory is of Gus walking through the schoolyard with half a dozen stray dogs barking and growling at him. Gus showed no fear toward the dogs. He had a long, heavy walking stick and he was swinging it at the dogs to keep them at bay.

One time while I was hunting with my .22 rifle and dog "Tip" at the very top of Silver Shield Gulch about a dozen goats came running up out of the brush and across a 20-yard opening straight toward us. They were obviously very scared and running away from something. I prepared to shoot whatever was chasing them. The goats veered off to my left and I never did see what was chasing them.

JAP CAMP

Jap Camp was a small community on the West Side of the hill between Fox's Switch and H-Level that was occupied by people of Japanese descent. There were about fourteen buildings that held thirty-five apartments in this camp; they were occupied mostly by single men and only a few families. The Miyagishima (Miya) and Kawaguchi kids were the only ones near my age. The Miyas ran a boarding house at which the single men ate. During World War II quite a few Puerto Ricans and Navajo Indians were brought to Bingham to work at the open-pit mine. Most of the Puerto Ricans lived in Greek Camp but they ate their meals and picked up their sack lunches at the Miya's boarding house in Jap Camp. Amy Miya Gonzales remembers putting up two hundred sack lunches every night for all the men who boarded there. (There were also one or two boarding houses down in Frog Town where many of the Indians and some Puerto Ricans lived.) Joe Miya was the Camp Boss and also worked part time for the Utah Copper Co. Only a few of the apartments had bathing facilities or even running water. The camp shower house was primarily for the single men but the families used it during certain hours. It provided some surprises to kids who had to go in there for a pit stop. Some of the houses in Dinkeyville and Telegraph didn't have bathtubs so a few of the boys sometimes used the showers in Jap Camp (they were called "strays"). On the south side of the tiny community there was a communal vegetable garden that was almost hidden among the trees. Kids used to sneak up through the trees and stretch their arms through the fence to swipe carrots.

Figure 85 – Jap Camp in 1926 Before Construction of the Shower House
Used by Permission, Utah State Historical Society, all rights reserved – Jap Camp encompassed about fourteen wooden buildings that were above Fox's Switch and below H-Level on the West Side of the canyon. Most of the buildings were apartments for single men but there were a few families. Miya's boarding house was the uppermost long building near the center of the photo.

Toilet facilities were limited at the time of this photo. A two-person outhouse was located on the near side of the road just past the right end of Miya's boarding house; there may also have been several other outhouses. The few families in the Camp had running water with sinks and toilets but not bathtubs or showers and they did their own cooking. The bathing facility consisted of a huge tub on the left (south) end of the boarding house that was used by the community – men right after work, families in the evening. The apartments for the single men didn't have running water so they used the outhouses and the huge tub and ate their meals at the boarding house. Most of the men also kept a bucket of water and a wash basin in their apartments.

An item of special interest was the pipe line that provided compressed air to the U.S. Company's mines three miles up the canyon in Galena Gulch (the U.S.). The air line came from the U.S.'s compressor behind the U.S. Hotel and followed along the ground from the lower right of the photo upward to the left side of the very long building on the right. To cross Jap Camp, the air line was held aloft by five wooden support structures about twenty feet high that angled upward across the camp; they were especially obvious where the line crossed the road in front of Miya's boarding house. The air line then descended to the ground and continued around the hillside and off the left side of the photo about twenty feet below H-Level where the ore cars can be seen.

The road to Jap Camp came from the left side of the photo; the lower branch went down to the bottom of the camp and then around, above the railroad tracks and above the U.S.'s change/dry house, to Greek Camp. This lower road was very narrow and dropped steeply down to the railroad

tracks below; at least one automobile slipped off the road and threatened to roll down onto the tracks. The upper branch of the road went slightly uphill in front of the boarding house and all the way to the right side of the photo where it ended as a small parking area from which a trail led up to the mine levels. There was a trash dump on the right (north) side of this photo by the parking area; we kids were awed by and curious about the small wooden barrels with Japanese writing that had been discarded there. The barrels, which had contained exotic sea foods, represented a foreign culture – they had been shipped all the way from Japan!

Figure 86 – Jap Camp Immediately After the Shower House Was Built, 10-17-1926
From the Gordon Bodily Collection –The shower house was the light-colored, taller, narrower building which sits at an angle and is right of the center of the photo. It contained showers, toilets, and a big multi-person sink about 3 feet by 6 feet. The shower house was built on the site of the two-person outhouse shown in the previous photo. Another photo taken 9-20-1926 may have showed some construction activity at the site but the shower house was not evident. The shower house must have been built very rapidly in September and October of 1926. It replaced all of the outdoor toilets and also provided showers and a large sink. However, the big bath tub on the south end of Miyagishima's (Miya's) boarding house continued in use until about 1943. Jap Camp, excepting for the shower house, was built before 1915 but a more exact date is not available.

Miya's boarding house was the long building on the upper road just to the left of the shower house. The air line that ran from the compressor behind the U.S. Hotel to the mines in Galena Gulch was elevated at this point; it crossed the roadway in front of the boarding house. The air line and supports were removed in the 1930s or early 1940s.

The roof of the school house, with the addition on the back, is in the foreground. This photo also shows the waste dumps above Jap Camp. They were built shortly after the open-cut mine started in 1906; waste was hauled to fill in Copper Center Gulch which is to the right of Jap Camp and to fill in much of Galena Gulch (the U.S.) to the left of and above this photo.

GREEK CAMP

Greek Camp was above the business district of Main Street. It had half a dozen two-story apartment buildings in which single men lived. One building was the dining hall where "Dan the Chinese Cook" reigned. There also was a shower building, with individual lockers, on the upper side of the camp next to the tracks where the men changed clothes and showered after work. Some of the men from Dinkeyville who worked for the Utah Copper also made use of this shower building. A lot of the boys from Dinkeyville went over there to shower because they had no showers or bathtubs at home. Besides that the water was always nice and warm. There were no families or children in Greek Camp so we kids had little reason to go there. My brother Garland, however, had a couple of deals with Dan the Cook. Gar got a porcupine on two occasions and took them to Dan who butchered them and made stews out of them. He split the stew with Gar who brought his half home. I can remember eating some and it was very good. Another thing I can barely remember is going over to the dining hall to see Dan the Cook who frightened me with a butcher knife and ended up giving me a big piece of apple pie. During World War II most of the Puerto Ricans and some Navajo Indians lived in Greek Camp but they ate at Miya's in Jap Camp.

Figure 87 – Greek Camp, Utah Copper Mine & Niagara Tunnel About 1926

From the Gordon Bodily Collection - The Greek Camp change room/shower house is the uppermost building, in left center, light color. The other buildings were apartments and a dining hall. The Utah Copper appears to have some steam engines or steam shovels but with some towers for electrical wires on the level just above and to the right of Greek Camp. Electrically powered shovels were first used on the hill in 1925 and electrically powered engines were first used in 1926. This photo was probably taken during the transition from steam to electrical power. Several water tanks for watering the steam engines can be seen on the levels above Greek Camp. The portal to the U.S. Company's Niagara Tunnel was under the large, long building in the lower left of the photo (see "Up the Shaft After School". Large stacks of timbers are in the yard outside the portal; they would have been used as they are or cut to specification in the carpenter shop before taking them into the mine. The chimneys of the Terrace Heights two-story duplex homes are evident – they were built in 1925-1927. This photo was probably taken about 1926.

TELEGRAPH

Telegraph was along the main road about a mile above the vehicular tunnel. In the 1940s and 1950s there were two one-story fourplexes and, a little higher up, a third fourplex. Three individual houses were located a quarter to half a mile farther up the canyon. The three fourplexes, total twelve houses, were identical in size and floor plan. The houses were owned by the U.S. Smelting, Refining and Mining Co.. There were only about four families in Telegraph, one of which was our friends the Halversons. Lee was my brother Garland's age and we often went rabbit hunting together. Emil Bodmer and his mother lived near the upper end of Telegraph; their house was somewhat isolated and the trees and brush of the hillside came down behind and to the sides of it. Some kids said they saw the mother talking to the deer by the house (I never knew whether the deer talked back or not). We used to go hunting above Emil's house and he never bothered us. In later years, when he definitely was a game warden, I don't know what his reaction would have been if he'd have seen us with our .22s.

Telegraph also had a nice level area below the houses that was covered with soft sand. It was an ideal ball field. The kids from Dinkeyville and Copperfield used to go up there to play football and baseball. This "ball field" was the leveled-off tailings pile of a mill that had been closed down and demolished decades earlier.

Figure 88 – Telegraph Mine About 1910
Photo Courtesy of Kennecott Utah Copper - The bunk house is the large three or four-story building in the upper right. All of these buildings on the upper dump were demolished and replaced with two rows of company housing before the 1940s. All the structures of the lower mine were also gone before the 1940s. The long snowshed-like structure on the left half of the lower dump could have been the covering over the Holden Tramway. A stamp mill that crushed and ground ore to the consistency of fine-grained sand was built and operated on the lower mine dump site but it isn't known if it was built before or after the mine structures were removed. All the structures of the mill were also gone long before the 1940s. The dump was leveled off and was covered with the soft sand from the mill. The nearby "Cocoa Dirt" was the even finer-grained reddish brown tailings perhaps from this same mill that apparently treated oxidized gold ores.

THE U.S.

The community of The U.S. was two miles or more above Telegraph, three miles above the Copperfield School. Twelve or thirteen houses were situated on two levels and there were a boarding house, mine office, assay office, and even a showhouse and clubhouse. Half a dozen kids from the U.S. went to the Copperfield School and one older boy went to Bingham High School in Copperton. Most of them were older than me. Silver Shield was a mine about half a mile below the U.S. but only a few children lived there and I didn't know them. It and three or four other scattered houses were collectively included in the U.S.. The U.S. used to be called the Old Galena and it was of special historical significance because, in 1863, it was the site of the first registered mining location in Utah. The houses and other buildings in the U.S. were abandoned and destroyed in about 1943.

The Silver Shield (Porcupine Gulch) also was of historical significance because there was, on its upper slopes, a grove of huge red pine trees (Douglas fir) 3 feet or more in diameter. These trees were called the "Big Grove" and they were used to build the roof of the tabernacle in Salt Lake City in the 1860s. The need for timbers to serve as ground supports in the underground mines only a few years later essentially denuded the area for miles around of all commercial-size timber.

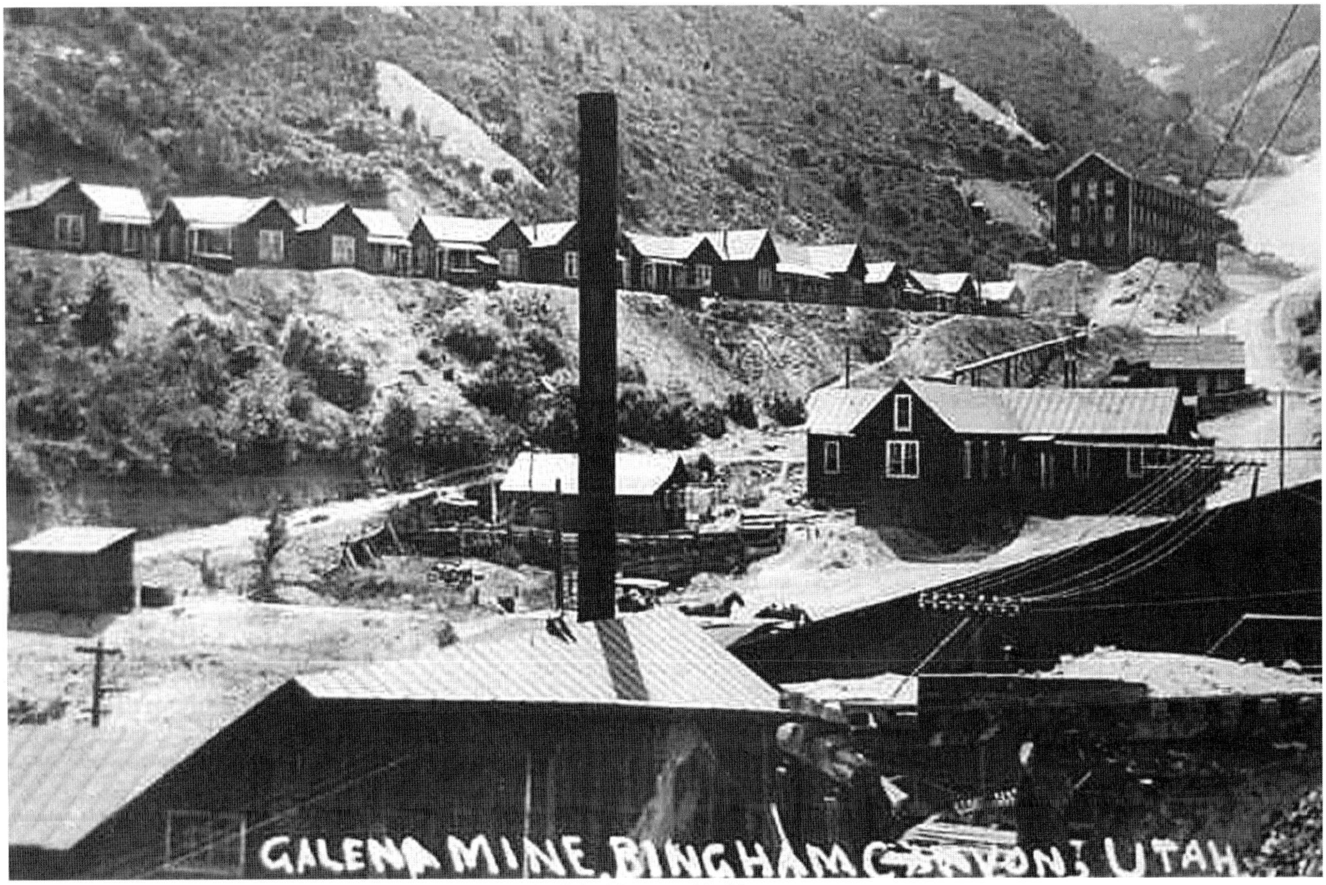

Figure 89 – The Old Galena Mine (The U.S.) About 1940
From The Gordon Bodily Collection - The Galena Mine was, in its later years, called the U.S. because it was owned by the United States Smelting, Refining, and Mining Company. There were houses for families, a boarding house and other buildings necessary for the mining operation. This photo shows some of the last remaining residences at The Old U.S.. Lorraine Clinton's house was just to the left of the smokestack (see "Up The Shaft After School").

BUSINESSES IN COPPERFIELD

Essentially all the businesses in Copperfield were located along the lower third of Main Street. Company houses occupied most of the upper two-thirds. Prior to 1945 there were several businesses including the U.S. Noodle House that I was too young to recall and several others such as the Miner's Merc, the Independent Grocery and the Dianna movie theater that I barely recall. In 1945 the businesses consisted of three beer joints, two or three groceries, a meat market, two tiny restaurants (including the Combination), a barber shop and a couple of apartment buildings. We kids would go into the beer joints to sell empty soda pop or beer bottles or to play the slot machines. The grocery stores would deliver orders to the people's houses and would also charge goods to an account. The grocery stores were owned by Greeks as were most of the beer joints. We kids went to the grocery stores to pick up orders for our parents or to buy candy for ourselves. One grocer in particular suspected that we would steal candy from him and watched us like a hawk. I don't doubt that some kids might have stolen a very small amount of candy from him but he treated all of us like we were wholesale thieves. We preferred to buy candy from the other stores but eventually the other two stores closed and we had to patronize the ever-vigilant suspicious grocer.

The meat market was owned by "Louie the Butcher" and was sometimes a source of income for enterprising boys. Louie would pay us to sack up and remove the old sawdust from the floor of his shop and to bring fresh sawdust from the carpenter shop of the U.S. Mine. We would also get scrap bones and meat from Louie to feed to our dog Tip. The tiny restaurant, called "Frenchy's" at that time had about eight stools and a couple of small tables. We would go to Frenchy's to buy a hamburger for lunch or to play the slot machine. Joe Berger also had a couple of stools that served as a restaurant in his Combination bar/grill but we kids never ate there. The apartment buildings were occupied by the business owners and some single men. Some of the Greek and Italian men had families back in the old country so it was unclear which of them were real bachelors and which weren't.

Fresh milk was delivered early in the morning door-to-door by Hogan's Dairy or by the R-Dairy both of whose dairies were down in the valley. Most people had a milk box on their front porch; the dairy would retrieve the empty bottles and replace them with full bottles on a twice-a-week schedule. The milk wasn't homogenized in those days so the cream would rise to the top of the bottle. In the winter when it was cold the milk often froze and expanded so that it pushed the cream and the cap about two inches out from the bottle.

One activity associated with Hogan's Dairy was that in the summer we sometimes thumbed our way down to West Jordan and went swimming (nude) in the canal about half a mile from the dairy. Currants grew along the canal and we usually ate a lot of those. After the swim we would walk back to Hogan's Dairy and buy some chocolate milk or orange drink and swig it straight from the bottle right there. Then we would thumb our way back to Copperfield.

Does anyone have a good photo of the Copperfield Merc which occupied the building before the Miner's Merc? Also, when did the Copperfield Merc become the Miner's Merc and when did the Miner's Merc close?

325 Miner's Mercantile Company Charge Bill

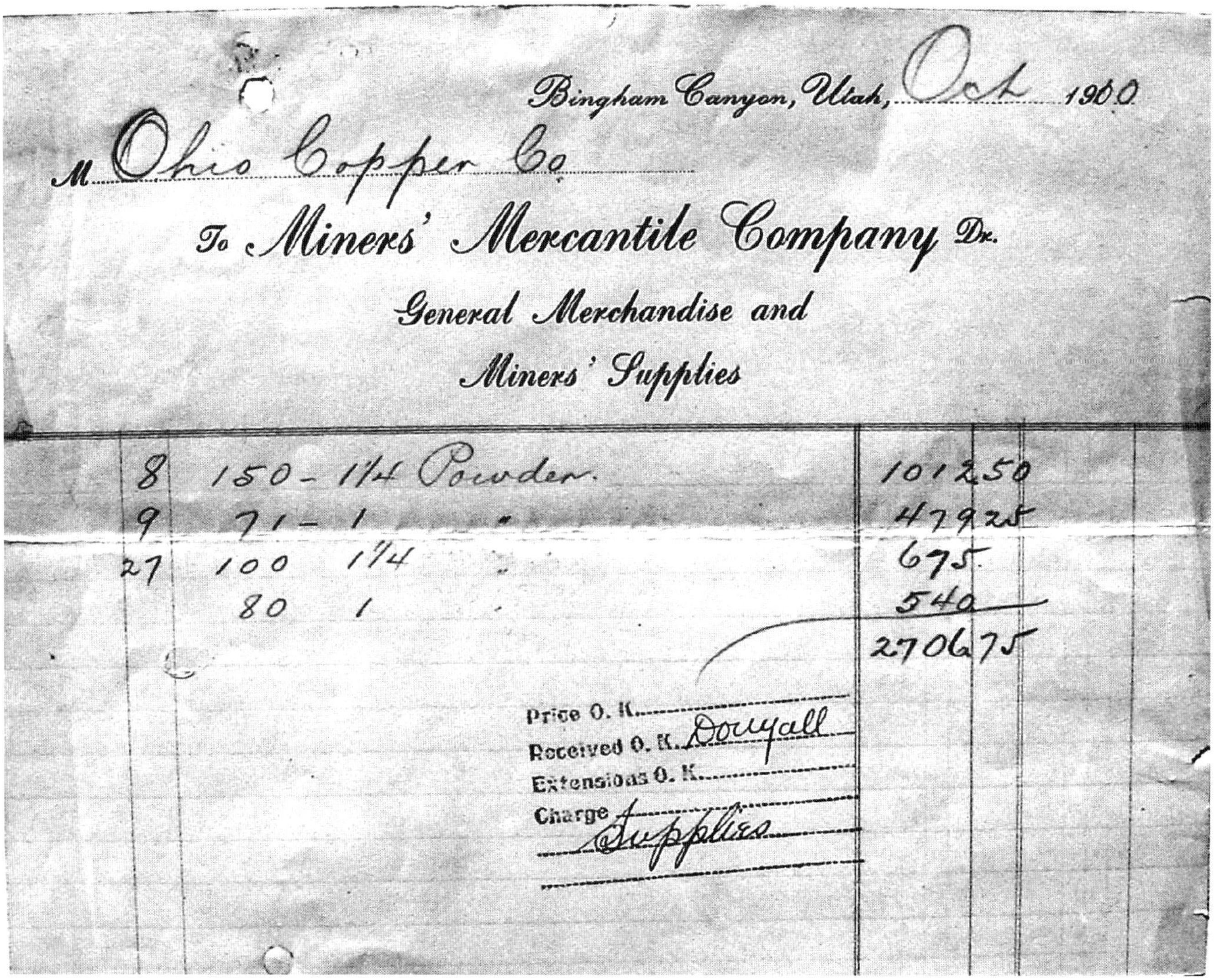

Bingham Canyon, Utah, Oct 1900

M Ohio Copper Co

To Miners' Mercantile Company Dr.

General Merchandise and

Miners' Supplies

8	150 - 1 1/4 Powder	1012 50
9	71 - 1 "	479 25
27	100 1 1/4 "	675
	80 1 "	540
		2706 75

Price O. K.

Received O. K. Dougall

Extensions O. K.

Charge Supplies

Figure 90 – Charge Bill of Miners' Merc to Ohio Copper Oct 1900
From the Gordon Bodily Collection – This charge bill predates the open-cut mining and is one of the earliest mementoes or indications of both the Miners' Merc and the Ohio Copper Company.

Another business in Copperfield was that of Oren Tolman who sold fruits and vegetables door-to-door. He lived in the valley and would drive his truck loaded with vegetables up to Copperfield once a week. He would hire several boys who would ride in the truck to a spot where he would park. Then the boys would fan out to all the houses and take orders for his goods. They would return to the truck with an order, put the fruits and vegetables into a sack, and deliver it to the house. The people would pay the boys who would then give the money to Oren.

One advantage of helping Tolman sell fruits and vegetables was that a kid could eat some of the fruit while riding in the back of the truck. This practice was readily apparent when Garland had an accident. He had worked for Tolman a good part of the day and then went down to the schoolgrounds to play. The boys were playing tackle football. Leon Miller tackled Garland and when Gar fell he broke his arm. They took him down to Bingham where Doctor Richards put him under an anesthetic and set his arm. While under the anesthetic he vomited. It amazed the doctor to see all those grapes!

Figure 91 – East Side of Lower Main Street, 4th of July Celebration 1936
Photo by Vera Johnson, From the Shirl and Isabell Scroggin and Eugene H. Halverson Collections – Businesses in this photo include the Combination Bar of Dick Hyland and Joe Berger, Mike's Barber Shop, the Miner's Merc and the Byrnes Hotel. Copperfield used to hold its own celebrations until about 1939. After that, as the number of businesses and the population of Copperfield diminished (and the length of Main Street was shortened) by the open-pit mine expansion, the celebrations were combined with and held in the main part of Bingham.

From "Down Memory Lane, Bingham In 1911" by John J. Creedon 30 Sep 1960 (excerpts) Courtesy of Gordon Bodily

To care for the horses and stable, there were several livery stables. The Bingham Livery was one of the largest. It was owned by George Chandler and was situated where the Princess Theater now stands. How those skinners used to come down that runway and make the turn up or down the canyon on two wheels was a thrilling sight to see. Some of the others were the Utah Livery, Star Livery near No. 2 Fire Hall, Blacks' near the Rio Grande Depot and the Bingham Coal in Hegland Ave.

Most of these stables kept saddle horses for hire and on Sundays and holidays every horse in town would be rented. These horses were trained to return to their stalls whenever they were turned loose. A man could hire a horse to go to Upper Bingham or Highland Boy and on arriving at his destination he would place the reins over the saddle horn and the horse would go home. It was very difficult to catch one of these horses for a free ride.

Figure 92 – West Side of Lower Main Street, 4th of July Celebration 1936
Photo by Vera Johnson, From the Shirl and Isabell Scroggin and Eugene H. Halverson Collections. Businesses in this photo include Louie's Meat Market on the left and the U.S. Noodle House on the extreme right. Individuals who are noted on the photo include Velma Johnson, who was Vera Johnson's daughter, Lorraine Clinton who was Vera's granddaughter (see "Up the Shaft After School") and Willimay McDonald who helped to make this photo available. Other persons, identified by number are (1) Eleanor McDonald (Willimay's mother), (2) young John Pantalone, (3) Mrs. Compagno and baby Phil, (4) Edith Leatherwood and (5) Gladys Scroggin (Shirl's mother).

From "Down Memory Lane, Bingham In 1911" by John J. Creedon 30 Sep 1960 (excerpts) Courtesy of Gordon Bodily

Bingham in 1911 was a typical western mining town. Where there were sidewalks they were wooden, about three feet wide and made of three inch planking.

The road was dirt and in wet weather it was a quagmire of mud. There was always a big mud hole by No. 2 Fire Hall, another at the top of Butcher Hill, about where the Post Office now stands and the biggest of all by the Bingham Merc.

When a six or eight horse ore team hit the mud, everyone scurried for a doorway or between buildings to get out of the line of fire. You could tell the amount of rainfall by the height of the mud line on the poles and buildings.

Ore teams with their heavy loads were constantly on the move. They used a heavy chain or rough lock on the wheels as an aid to their brakes. In the winter the wheels were replaced by sleigh runners and rough locks were used on these also.

Figure 93 - U.S. Noodle House on Main Street Copperfield July 4, 1936
Photo by Vera Johnson, From the Shirl and Isabell Scroggin and Eugene H. Halverson Collections - The U.S. Noodle House was on the west side of the street across from Joe Berger's Combination Bar. Louie's Meat Market was immediately to the left of the Noodle House and the Cairo Club (a coffee shop) is shown to the right. This photo was taken during a Fourth of July celebration. The U.S. Noodle House closed its doors in about 1938; Louie's Meat Market remained in business until the mid 1940s. Individuals who are identified by number in this photo include (1) Gladys Scroggin, (2) Eva Watson, (3) Ray Watson, (4) Thelma Barrett, (5) Stella Klopenstein, (6) Mary Kokla Zaccaria, (7) Blanche Leatherwood, (8) Edith Leatherwood, (9) Alice Ivie Carrigan, (10) brother of Alice (Keith), and (11) Hazel Skinner.

Figure 94 – Pan-Hellenic Grocery, Combination Bar, Mike's Barber Shop About 1957
Photo Courtesy of Shirl and Isabell Scroggin - The mine at this time is threatening to engulf these buildings. The Pan-Hellenic Grocery is nearest the mine, next to the pickup truck. Joe Berger's Combination Bar is the center of the three store fronts with the "Entrance" sign. The location of Mike's Barber Shop was the third of the three store fronts, on the right – it was probably closed at this time. The Byrnes Apartments were on the second floor, above the stores. The Pan-Hellenic Grocery was the last grocery store in Copperfield; the Independent and Lendaris Merc groceries had closed several years earlier. Joe Berger's "camper" is parked on the lot vacated by the Miner's Merc; the Independent Grocery was adjacent to it.

Joe Berger's (with partner Dick Hyland in early years) Combination Bar/Restaurant was well known in Bingham and an article about it was printed in the Salt Lake papers. There was a counter immediately inside the door where candy and soda pop could be sold to kids. This 5-foot space was separated from the bar area by a small divider which few people paid attention to. There were six or eight stools in the bar itself of which two at the far end were for his "restaurant". An area behind the bar/restaurant was closed off by swinging doors; there were two pool tables there. Joe had a retinue of special customers with or for whom he had parties from time to time. Some of these special occasions included his cooking such things as fried grasshoppers. The occasion that I recall was the rattlesnakes. He bought some big rattlesnakes and put them into a glass-topped display case or cage with sand on the bottom. The snakes were kept in the case for several weeks and he fed them whole raw chicken eggs and maybe other things. I can remember the eggs as having "grooves" in the shells that the snakes had somehow made. I don't know whether the snakes ever actually swallowed an egg or not. Joe had pictures taken of him in his high-top white chef's hat butchering and cooking the rattlesnakes. Another entrée for Joe's restaurant was porcupine stew. He would pay kids fifty cents for each porcupine they brought in to him, unskinned. Skinning a porcupine could be a tough job so the kids, luckily, didn't have to do that – Joe did it himself. My Dad was a customer of Joe's and my brother Garland, as a young boy, worked for Joe in cleaning up the place and removing clinkers from the furnace.

Mike the Barber may have closed shop several years before this photo was taken. The unusual thing I remember about him is that he had an open cabinet in which he placed his equipment and supplies. One part of the cabinet had a stack of towels. It was said that he kept a loaded revolver hidden among the towels and that he was ready to use it if required. It was also said that years ago some men had tried to rob a business across the street and Mike had used his revolver to help stop them. While getting my hair cut I would look at that stack of towels and try to spot Mike's revolver but it must have been pretty well hidden because I never did see it. I was also very careful not to offend Mike.

COPPERFIELD GRADE SCHOOL

The original Upper Bingham (Copperfield) Grade School opened in 1909, the same year as the school in Lark. It was a 2-story building with the heating plant and bathrooms inside the school. It cost $18,000 to build. The school was located on the east side of Main Street about 500 feet above where the entrance to the vehicular tunnel would be built thirty years later. It had been open for only nine years when, in July of 1918, it was destroyed along with the Caulfield and one other apartment house and three homes by a devastating fire.

After the first school burned down a new two-story school was built on the same site as the old one and opened in 1919. The heating plant and bathrooms for the school were housed in an immediately adjacent building on the north side of the school.

In 1926 a flat-roofed two-story addition was built onto the back of this second school. The first story was an auditorium/gymnasium with a stage; the upper story had two classrooms (5th and 6th Grades) and a library. Behind the addition, on a level with its roof, there was a dirt (later black-topped) parking lot. The parking lot was essentially built on part of the old Shay Track. An open "alley", about four feet wide, separated the building addition from the supporting wall of the parking lot.

The entrance to the school and the school yard and playground were elevated about six feet above the roadway. The north part of the schoolyard, or playground, had half a dozen swings or rings suspended from chains attached to a support that was made from six-inch iron pipes. There also was a vertical pole with a movable "head" to which about eight rings were attached by one-inch chains; kids could hold onto the rings and run in a circle as the "head" rotated. We called it "The Ring". By running fast enough and then jumping upward and outward you could leave the ground and fly through the air for half a rotation or so. If enough kids were playing on the "ring" and they were running fast enough, one or two kids, especially small ones, could get a free ride and hardly touch the ground as the centrifugal force pushed them outward and kept them whirling in the air. The south side of the schoolyard had a metal slippery-slide and a teeter-totter. It was also the location of the flagpole.

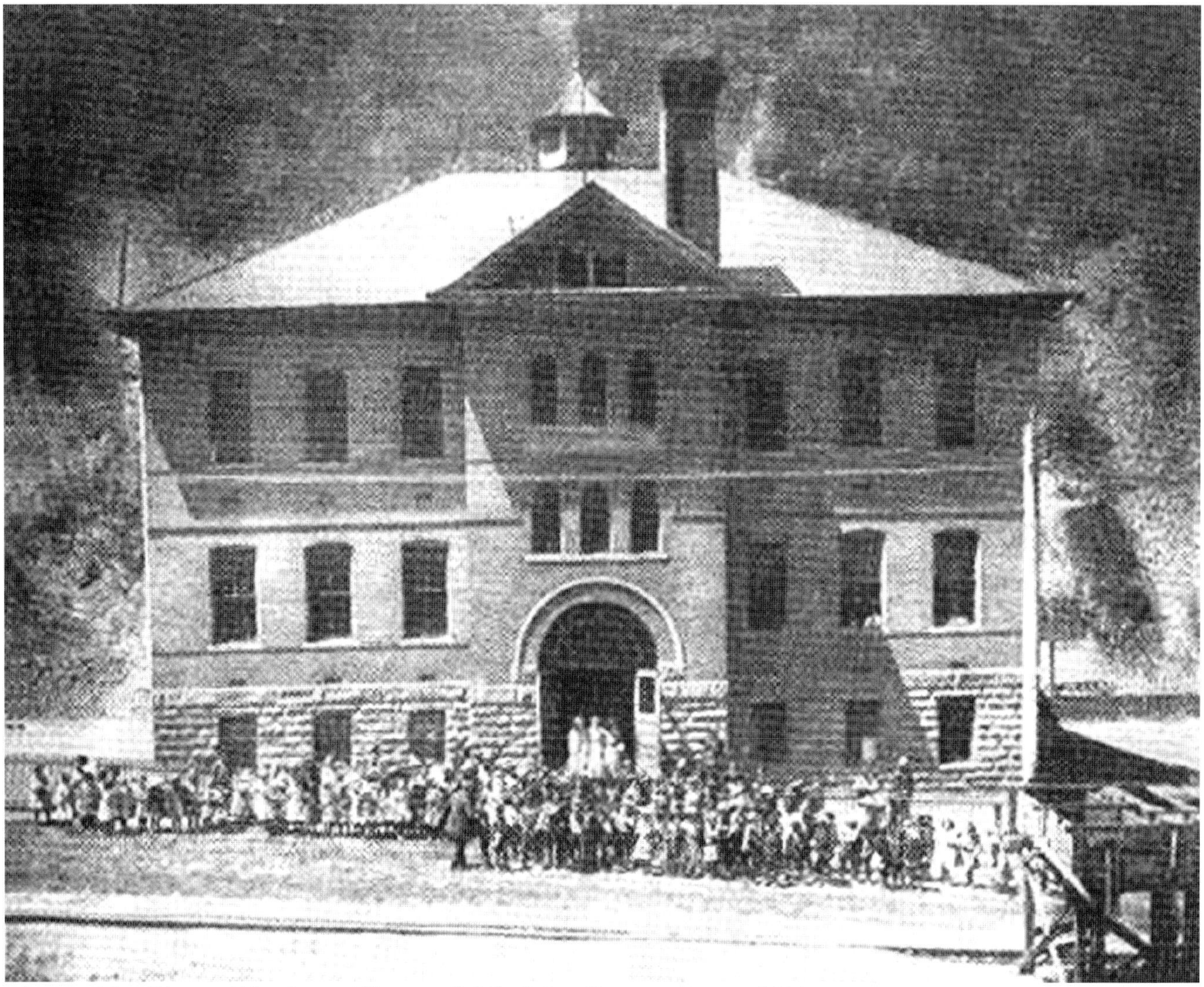

Figure 95 – Upper Bingham (Copperfield) Schoolhouse About 1909
From "Bingham, The Greatest Copper Camp In The World" by The Bingham Commercial Club, 1909 - This first school was built in 1908 and opened in 1909. It was referred to as a two-story, three-room schoolhouse but appears to have more than three rooms. The heating plant and bathrooms were indoors. The chimney for the heating plant is on the roof to the right of the door. No other buildings are evident to the sides or behind the school. Terrace Heights had not yet been built. The schoolyard appears to join directly onto the roadway without a fence or any other obstruction. There are about seventy or eighty students and their teachers in this photo.

Figure 96 – 1924 Photo of Copperfield Grade School (Rebuilt in 1919) Before the Addition Photo Courtesy of Kennecott Utah Copper – There is no addition behind the main building. The building that housed the heating plant and bathrooms is on the other side of the school building. The front entry of the school lets straight out and down onto the roadway below. The Company houses along Main Street have been built; Leatherwood's last house was on the same side of the street as the school – it is in the extreme left of the photo. The two houses to the left of the school, slightly above but behind those on Main Street were occupied by Barretts and Scroggins. The one-story duplex Terrace Heights houses are under construction in the upper right. Construction of the two-story Terrace Heights houses had not yet started (they would be built in 1925-1927).

The first floor of the school had the First Grade and Second Grade classrooms and the auditorium with a stage at its south end. The second floor of the school had classrooms for the Third, Fourth, Fifth, and Sixth Grades and the library and administrative office.

Kids from grades one through six went to school in Copperfield. For grades seven though twelve we went to the Bingham High School in Copperton. When I went to school there the Principal was Mr. Atwood and the Librarian/Custodian was Blanche Leatherwood. There was one teacher for each of the six classes but there was a fairly high turnover. Each class had about twenty-five students so there were about a hundred and fifty kids who attended school each year in Copperfield in the 1940s.

Figure 97 – Rebuilt 1919 Grade School About 1956 With 1926 Addition On Rear
Photo Courtesy of Shirl and Isabell Scroggin – This school was built in 1918-1919 and the addition of the auditorium, with the 5th and 6th Grade classrooms and the library was added to the back of the schoolhouse in 1926. The heating plant and bathrooms are to the left of the schoolhouse. The "alley" behind the addition can be seen. One side of this "alley" was formed by the outside wall of the addition and the other side was a concrete wall about thirty feet high with three support pillars. A protective fence separated the parking lot from the "alley" behind the school. The one-story Terrace Heights duplexes that were started in 1924 are behind the school. The Copperfield Grade School closed in 1958.

The school did a good job of teaching us how to read, write, and do arithmetic as well as the other skills that are normally taught in grade school. Our teachers were all dedicated and fair. I can still remember most of them. When we went to the Seventh Grade down in Copperton I believe we were well prepared for the transition.

We were a very mixed group of boys and girls of several nationalities and I don't think we mistreated anyone because of their race or background. We all played together on the school ground at a variety of games and had fun doing it. Yes, we sometimes had our differences just like at any other school. Some of our games were a bit rough – especially on the parking lot after school and we incurred our share of bruises, scratches, and scrapes.

My brother, Garland, still remembers a humorous incident when he was in the First Grade. He had a troublesome cowlick of hair that fell down over his forehead so before he left for school in the morning our mother would plaster his cowlick down with some kind of wave-set. One day Gar did especially well in a reading class and the teacher, Mrs. Hooten, patted him on the head for doing so well. Her hand came away covered with gooey wave-set! This was totally unexpected and very messy for her – she made Garland go to the lavatory and wash it out of his hair and she never patted him on the head again.

Figure 98 – The Copperfield Grade School After 1926
Photo Courtesy of Shirl and Isabell Scroggin - This is the rebuilt school, opened in 1919 after the first one burned down in 1918. The concrete retaining wall to the right of the school was built at the same time as the addition to the rear of the school. A corner of the bathrooms/heating plant can be seen to the left of the school. All of Terrace Heights has been completed – including the garages in the upper left corner. The Upper and Lower Terraces of two-story duplexes are to the left of this photo.

Another funny incident for Gar in grade school was when he had some difficulty in an English class. Two of the girls – Zenaida Aguayo who was of Mexican descent and Mary Kawaguchi who was of Japanese descent helped him with his English.

The "alley" behind the school house was a source of adventure. Most boys, at one time or another, climbed onto the fence at the parking lot and leaped across the alley onto the roof of the auditorium. We didn't do anything that was especially fun after we got onto the roof but did get a feeling of satisfaction from proving that we were daring and could do it. One girl, Dorothy Petrarca, in an attempt to jump across the "alley", fell the full thirty feet to the ground and was badly injured.

Figure 99 – First Grade Class of Copperfield School With Mrs. Hooten in 1937
Photo Courtesy of Lorraine Clinton Beckstrand – Mrs. Hooten taught the first grade in Copperfield for several years. Thirty-two students are present in this photo. Through the forty-nine class years from 1909 to 1957 there would have been over a thousand students who received their early education in the Copperfield School.

An incident that I still remember was when Garland and I were playing on the big iron ring that hung down next to the swings. There were two or three eight-inch rocks right under the ring and there was a larger rock fifteen feet or so off to one side. Garland was hanging from the ring and swinging back and forth. He was about twelve years old and I was only eight. We decided that when Gar swung over by the big rock then I would leap from it and cling to his back. He swung over fairly close to me but I was afraid to leap that far. Since I hadn't jumped when I should have he stopped paying any attention to me. But when he swung near me the next time I leaped and wrapped my arms and legs around his back. Gar wasn't prepared to hold both of our weights and he lost his grip on the ring. We both fell, with me underneath. I landed on my back right on top of an eight-inch rock with Gar on top of me. My backbone took the full impact and I can remember lying there and not able to move my legs! They were definitely paralyzed for a few moments. Then I was able to get up. Twenty years later I had an x-ray taken and the doctor noted that my back had been injured at some time in the past.

The front schoolyard was where I fractured my skull while roller skating. It was also the site of another odd type of entertainment -- flagpole swinging. This was a form of amusement that I witnessed but never tried myself – it just looked too risky or foolhardy to me. The flagpole in the front schoolyard was about thirty feet high and was made from a 6-inch iron pipe. Some of the more daring, older boys would get hold of the rope for the flag, climb onto the slippery slide and then launch themselves like Tarzan swinging from a vine in the jungle. They would deliberately launch themselves outward toward the roadway. They would swing clear out over the top of the fence and over the sidewalk and the edge of the roadway below. The flagpole would bend to the side under their weight and momentum. I fully expected the flagpole to snap and could imagine somebody getting killed but the flagpole never broke.

One comical incident was in 1941 when Harold Nielsen came to the Copperfield Grade School in midyear to serve as the Principal and 6th Grade Teacher. One day he went to the furnace room and was astounded at an unexpected sight. The custodian, who lived in Dinkeyville, apparently didn't have any clotheslines at his house or the weather was too wet to dry their clothes outside. So he strung clotheslines throughout the furnace room and had his family's clothes hanging on the line right next to the furnace, the coal room, the stoker full of coal and the clinkers. The Principal never expected to see freshly laundered clothes in such a precarious location. One slip and the clean clothes would have been blackened by the coal or soot. He laughed about it for years afterward.

THE U.S. HOTEL

The U.S. Hotel was a 4-story building located on Main Street across and a little above the vehicular tunnel. It was the largest building above the main business district. The bottom story contained some offices, the second story had a large lobby, small commissary, a dining hall, kitchen, and living quarters for the manager. The top two stories each had about twenty six rooms for the workers and a community bath. A total of about fifty-two workers for the U.S. Mine lived at the U.S. Hotel. A flight of stairs attached to the outside of the building ran from the ground level up to the covered porch on the second floor where the front entrance to the lobby was located. A back door from the lobby led up to the yard of the mine.

The U.S. Hotel was significant to the kids who caught the school bus to go to school down in Copperton. The lobby and the area under the porch provided cover and warmth during cold weather while they were waiting for the bus.

One of the funny incidents I witnessed was when, on a particularly snowy morning, the stairs leading up to the lobby were so covered and filled in with snow that the stairway resembled a slide with only a small gap between the steps. One heavy-set boy who wasn't very athletic made it up the stairs but on his way back down he fell near the top step. He slid from one step down to the next with a thump and jolt and then on to the next and the next, unable to stop his downward movement. The result was a "thump-thump-thump" as he slid all the way to the bottom of the stairs! It was like something out of a comic strip! His rump must have been bruised at the end of this mishap.

The hotel was also important to me because I could warm up there when delivering newspapers. They had some steam heaters in the lobby and I would lay my wet gloves on a heater to warm up and dry out while I delivered papers to the individual rooms on the upper floors.

Figure 100 – U.S. Hotel, Air Compressor, the Niagara Tunnel and Ore Bins In Early 1950s
Photo Courtesy of Kennecott Utah Copper - The U.S. Hotel had, as indicated by the windows, 14 rooms on this side of the building on each of the top two floors. On the other side there was a bathroom and 12 rooms for individual men on each of the two floors for a total of 52 rooms. The smokestacks for the air compressor/heating plant are right behind the hotel. The compressor provided pressurized air to power the drills inside the mine and for the high-pressure air locomotives that were used on the main (Niagara) level. Compressed air was, for some years, also piped up to the mines in Galena Gulch (The Upper U.S.). The portal of the Niagara Tunnel of the U.S. Mine is directly beneath the large Change/Shower House building that is in the top center of the photo. The ore bins for the Niagara Tunnel are in the right side of the photo. The ore from the U.S. Mine (Niagara Tunnel) was hauled by small locomotives and ore cars (see the photos in "Up the Shaft After School" out from the tunnel and around to the ore-dumping facility in the center of the photo. Large standard-gauge railroad cars were on the tracks below the structure and the ore from the small cars on top the structure was dumped down into them. The ore chute to the right of the hotel was built after the hotel and after 1926. The model years of autos were used to date this photo.

Does anybody know when the ore chute next to the U.S. Hotel was built and how it was used? What were the circumstances of its construction and use? Was ore loaded into trucks and hauled away to the concentrator in Midvale? Was the ore chute only used when the Utah Copper was on strike and not able to haul the ore cars away?

THE MINE OBSERVATION POINT

The observation point for the open-pit copper mine was an important part of the Copperfield community. It was located at the end of the road about a quarter mile below the vehicular tunnel. The observation point stood on the very edge of the pit where it commanded a magnificent view of the entire mine. It was large enough to shelter several hundred people at a time. It was covered by a roof of heavy boards for protection against the weather and flying rocks and the ground was blacktopped so that it was never muddy. During the summer tourist season dozens or even hundreds of automobiles and half a dozen Greyline buses full of tourists would come to the observation point each day to look at the mine.

Kennecott had a man assigned to the observation point who would help the tourists and give a spiel about the mine – its history and statistics about the equipment and mining methods mostly. Scotty Robertson did this for several years and was greatly appreciated by the visitors. Scotty always wore a plaid tie and Kennecott received letters of appreciation from all over the world that were addressed to "The man in the plaid tie".

The tourists, to get to the observation point, first had to run a gauntlet of narrow streets and snarled traffic just to get to the lower entrance of the vehicular tunnel. Those who drove their own cars often had to carefully maneuver past other cars and obstacles with only a few inches of clearance. Some of them almost sweat blood during this trial of their driving skills. The buses had an even more difficult time getting up the main canyon but the passengers could at least sit back and let the driver do it all. After making it to the lower entrance to the tunnel the cars and buses usually found that the traffic light was red so they had to wait fifteen minutes or so for the downhill traffic to exit. Then they entered the mile-and-a-quarter long tunnel which itself was an adventure for many of them.

When they exited the upper end of the tunnel the tourists were often beset by teenage or younger boys (and girls) who would jump onto the running boards or run alongside the cars and offer to guide them to a better view of the copper mine. If they accepted the offer then they would be guided to some outstanding views on the higher levels that weren't being mined at that time. Those who chose to go to the "regular" observation point would turn right and drive down to the parking area at the end of the road from where it was a walk of fifty yards to the observation point where they could gaze at the huge open pit and listen to the description provided by the Kennecott representative.

The prime time to view the mine was about 4 pm because the blasting was conducted then (after the regular dayshift crews had ended their shift). Huge blasts of tons of dynamite would break the rock so that the electric shovels could dig it and load it into railroad cars to be hauled to the concentrators in Magna. The blasts would be preceded by the piping of warning whistles. Everybody would turn to look in the direction of the whistles and anxiously await the explosions. The Greyline buses generally scheduled their tours to coincide with the blasting.

From the time they arrived at the parking lot the tourists would be approached by boys and girls who would offer to sell them ore specimens from the mine. Some of the specimens really did come from the open-pit mine but most of them came from the ore bins of the underground U.S. Mine. Not only did most of the specimens not come from open-pit mine but some of them rarely occurred there. In later years an "official" purveyor of souvenirs had a small stand adjacent to the observation point. The owner of the souvenir stand was at a disadvantage as far as mineral specimens were concerned because he wasn't able to filch from the ore bins of the U.S. Mine like the kids did. The souvenir stand did, however, carry items that the kids didn't – like postcards and machine-made copper items such as pins with a tiny copper shovel upon which a very small mineral specimen was wired or glued.

Figure 101 – Mine Observation Point With Greyline Tour Buses About 1944

From the Gordon Bodily Collection - There was very limited space for vehicles to park, turn around, or even to drive. The line of cars at the right, including Joe Berger's "camper" and the two cars on the left are parked. The buses, with their loads of tourists, are threading their way back up toward the tunnel. The Observation Point, at top left, isn't crowded now without the three busloads of tourists. The open-pit mine is at the top center of the photo; the Observation Point is literally on the edge of the mine. The tour buses generally scheduled their trips so that the daily blasting could be observed. Blasting was done after the regular day shift had ended and before the evening shift started so that there were very few workers in the mine except for the powder gangs. Generally there were about eight blasts per day. Several minutes prior to the blast the powder gang would sound an air-powered warning whistle near the site of the coming blast. Tourists at the Observation Point could hear the whistles and watch for the next blast in the vicinity from which the whistle sounded. It was much like the 4th of July in many respects with whistles followed a few minutes later by blasts. Very often several whistles were tooting at the same time and their blasts went off within a minute or so of each other. The newest automobiles in this photo were early 1940s models and the buses were probably 1942 to 1946. The large two-story building on the right was occupied by the Pan-Hellenic Grocer, Joe Berger's Combination Bar, Mike's Barber Shop and, upstairs, the Byrnes Apartments. The vacant lot of the Bingham Merc is next and then the Independent Grocery is in the lower right of the photo.

The observation point was a part of life in Copperfield especially for the kids who sold ore there. The large amount of vehicle traffic that was generated by the tourists affected everyone, causing traffic jams and delays throughout the summer.

My brother Garland had the job for a couple of years of maintaining some coin-operated telescopes at the observation point. The tourists would insert a coin and for several minutes they could swivel the telescope around to get close-up views of especially interesting activities or machinery in the mine. The telescopes were owned by young Dave Curry.

PEARL DIVER GIVEN FLOATER

An amusing case was before Judge John C. Green Monday afternoon when Harry Davis, alias "Blinky," but who is better known as "Redlight Bill," was arraigned on a charge of vagrancy.

The defendant was arrested last week by Chief Pautsch and when his case was called he appeared for trial without counsel. When asked by the court the nature of his occupation he stated that he was a "pear liver." This answer puzzled the learned judge who desired more light on the subject. The defendant then explained that he was a "crockery massager", or to still further simplify the matter, he said that he was a common dish washer. But while that was his profession, it appeared that he had not been industriously devoted to that line of work during his sojourn in Bingham.

When his case was heard the judge informed him that hereafter he would have to feed in other pastures or spend a vacation of 90 days in jail. It was hard question for the defendant to determine. He has spent many very pleasant days in many different jails, but during the summer months there are places on the outside just as nice, so he decided that he would evacuate the camp. He was given 24 hours' time in which to wind up his business affairs and tell his friends good-bye.

The above article was printed in the May 24, 1918 edition of The Press-Bulletin, Bingham Canyon, Utah. Used by Permission, Utah State Historical Society, all rights reserved.

Figure 102 – Powder Monkeys Loading a Blast Hole

From the Gordon Bodily Collection - The blast holes were drilled in the toe of the bank to a depth of 30 feet. Then a few sticks of powder were pushed through a metal pipe to the bottom of the hole and were set off. This "springing" (small explosion) shot a quantity of rocks out through the hole like bullets from a rifle and created a chamber at the bottom of the hole into which successively larger charges of dynamite were inserted and set off until the chamber was large enough to hold the full amount for the final big blast. The powder monkeys had a difficult, dangerous job for several reasons. First, they had to climb down on ropes from the level above to scale-down loose rocks from the bank that otherwise might fall on them. Second, and perhaps the most dangerous, they used a long wooden pole to push the sticks of dynamite down through a metal pipe into the hole. Several powder monkeys were electrocuted when the metal pipe, as they pulled it out of the hole, came into contact with power lines for the locomotives or shovels. Third, they were working with explosives which can always be dangerous. There were instances of the stick or the pipe getting stuck in the hole after the fuse had been lit so that when the dynamite went off the stick or pipe came shooting out like a javelin and flew through the air for a record distance. This method of drilling and blasting was replaced in later years by the use of track-mounted rotary drills that drilled vertical holes down from the level above. Dynamite was largely replaced in later years by a mixture of ammonium nitrate and fuel oil. The powder box, to the right, was a very common but still coveted item by the inhabitants of Copperfield. The boxes served a myriad of uses. Most of them were used as kindling wood for building fires in the winter but they also were used as storage boxes and even nailed to the inside walls of houses to provide insulation and as wall coverings.

Figure 103 – Blasting of Ore
From the Gordon Bodily Collection - This was a small blast of only a few holes; each hole contained several boxes of dynamite. The shovel and the electric locomotive have pulled back from the site of the blast for safety purposes. Most blasts were several times larger than this and many were truly spectacular. Huge blasts were scheduled several days in advance and word spread around town about the upcoming event. Tourists, photographers, and even locals would gather and eagerly await the event.

THE GEMMELL CLUB

The Robert C. Gemmell Memorial Club was constructed in 1923-1924 by the Utah Copper Company and was opened in October of 1924. It supposedly was only for Company workers and their families. But nobody really cared if our Dads worked for the Utah Copper (Kennecott) or not – everybody was welcome to go there. The Gemmell Club was about 2 miles away from Copperfield down in Carr Fork. We (boys especially) went there quite often. The Club had a gymnasium, stage, and lounge on the main floor and a bowling alley, pool room, boxing ring, and showers in the basement. When there weren't any special events such as dances or boxing matches we would play basketball in the gym or we could bowl when there were lanes that the adults weren't using. Some boys went there especially to use the showers but very often we went to the Club just to mess around.

Before automobiles were popular and when almost everyone lived in Bingham rather than in the valley, special events were held in the Club more often. Boxing matches were a popular form of entertainment; a ring was moved from under the stage into the middle of the gymnasium. There were seats in the balcony above the gym floor as well as in the gym itself. Professional as well as amateurs fought in the Club.

Figure 104 – The Robert C. Gemmell Memorial Club In 1931
Used by Permission, Utah State Historical Society, all rights reserved - The Gemmell Club is the large building in the center. It was built in 1923-1924 as a memorial to R. C. Gemmell who was the General Manager of the Utah Copper at the time of his death in 1923. To the right is the tramway to the B & G yards and company offices. The 4-story building on the right is the Cyprus Hall boarding house. The Swedish Lutheran Church is directly across from the Club. The building to the left of the Club is a rooming house.

Figure 105 – "Smoker" or Prize Fight at the Gemmell Club
Photo Courtesy of Kennecott Utah Copper - The date of this fight or names of the fighters are unknown. The ring is in the gymnasium. Spectators are seated in the balcony as well as in the bleachers and on chairs on the floor of the gym. The ring was stored under the stage when not in use. There was another permanent ring in the basement.

SCHOOL BUS

Each school-day morning one school bus would pick up, at the upper portal of the tunnel, all of the Copperfield kids who went to the Bingham High School in Copperton. The same bus would deliver us back to Copperfield at the end of the day. There were enough kids to almost fill the bus. We got onto the bus about forty-five minutes before school started and spent another forty-five minutes getting back home after school. An hour and a half every school day was spent on the bus so it was a significant part of our lives. The bus service was reliable and everybody got along reasonably well. We could sit with our friends and we generally gravitated to the seats that we preferred – some up front, others to the back and everywhere in between. Over the years there were half a dozen different drivers. We were generally well behaved and rarely created problems for the drivers.

BINGHAM HIGH SCHOOL IN COPPERTON

The Bingham High School in Copperton accommodated all the kids in the Bingham-Copperton-Lark area from grades seven through twelve. There were about fifty-five kids in two classes for each grade; the total enrollment was about three hundred thirty. It was a bit of a shock to go from the smaller, local Copperfield Grade School where I knew almost everybody and was one of the older, larger kids and be suddenly thrust into such a large school where I knew only a fraction of the kids and was among the very smallest.

Figure 106 – Bingham High School in Copperton
Photo by Eldon Bray - The Bingham High School in Copperton opened in 1931; there were three other Bingham High Schools up in the canyon previously. This was the front entrance to the school; the Administration offices were on the ground floor to the left of the doors. The two wings of the building extended to the right and to the left of this photo. Another building behind this one contained the auto mechanics and shop areas. The gymnasium was a separate building farther to the left. Most of the students were bused from outlying areas to the parking, loading, and unloading area behind the school. A new Bingham High School opened in 1975 in South Jordan.

This was to be the major guiding influence in my life for the next six years. The number of teachers, with two classes for each of the six grades and a couple to spare, was about fourteen. With a turnover of about two each year there were probably about twenty-six teachers plus the principal, Tommy

McMullin, and support staff who helped to guide us students during my six years at Bingham High. Almost all the teachers maintained a positive environment for their students and for the entire school; some of them were outstanding. In general there were one or two subjects that each student especially enjoyed and was best at. The teachers naturally reciprocated by taking a greater interest in their most enthusiastic students who then gained much more than most kids did from the class. Several of the teachers had an uplifting influence on me.

There were many other influences in school that helped to shape us as adults. Other students had a strong effect on each of us. There were many good examples and there were a few bad ones just like in any other school. We had our choice of who to pal around with and who to follow. We influenced each other in all ways including scholastics, morals and social skills. I made some good friends in High School and value them to this day. Sports probably received more than its due share of emphasis. Unfortunately scholarship wasn't emphasized as much as it should have been with the result that most kids, including myself, coasted rather than working hard at our studies. Several of the science and math classes didn't have to be taken for graduation so we didn't take them and suffered for it in college. One positive incident that especially helped me in school was an aptitude test that was given to each of us in our Senior year. The results of this test suggested that I had a strong aptitude to be a Geologist and the more I thought about it the better it sounded.

WORKING ON THE TRACK GANG

All the boys at Bingham High School, between our Junior and Senior years, were given the opportunity by Kennecott to work on the track gang for the summer. Almost all of us were very glad to get a job because the pay was very good. After graduating from high school we could again work for Kennecott during the summer and then either continue on as full-time employees at the mine or go to college and work weekends, holidays and summers during our additional years of schooling.

When I graduated from the twelfth grade in 1952 I went to work for Kennecott for my second summer on the Nine-Gang on the East Side dumps. During that summer I worked with two young men from Midvale who encouraged me and helped me to register at the University of Utah. Without them I might not have started school that year or at least not on time. I was also able to pay my way through college by working on the track gang on weekends and summers – this was a great opportunity that Kennecott provided to college students.

At the University of Utah I found that in high school I had developed poor study habits and should have taken several more science and math classes. It took me close to a year to learn how to study and to catch up in some classes.

The Track Gang was the job at which essentially all employees of the Utah Copper (Kennecott) started. It required no special skills and was generally hard work. It was manual labor using a pick, shovel, spike mall and clawbar. A strong back was a definite asset. There were miles of railroad track in the mine and on the waste dumps. New track had to be built as the mine and the dumps expanded and all the tracks had to be moved and maintained. Straightening and tamping ties, driving spikes, mucking off the ties and leveling under them, stacking ties, lifting rails – this was the kind of work we did.

All that hard work might seem like punishment. But working on the track gang was a very good deal for high school and college students especially. We were paid the same as the men with families who worked alongside us and those were good wages for a 17-year-old kid. Hundreds or even thousands of young men earned the money to put themselves through college by working at the Utah Copper.

Some men worked full-time at the mine on the night shift and went to college part-time during the day. This arrangement was used by dozens of men – especially those who were married and had a family to support. However, by taking only 8 or 10 credits per quarter it might take them 9 or 10 years to get a Bachelor's Degree. After 3 or 4 years of working five nights a week and going to school three days per week it could be discouraging to realize that they still had five or six more years to go.

Then came the big layoff and it was the best thing that ever happened to a lot of these worker/students. Kennecott had to reduce their workforce for economic reasons and most of the students were laid off. At this point they had to decide whether to get another job or go to school fulltime. Most of them chose to go to school fulltime; they were rewarded by graduating from college much sooner and by obtaining jobs in their chosen profession at a higher pay than they had been earning at the mine.

Figure 107 – Track Shifter and Track Gang Working on Dump
From the Gordon Bodily Collection - The track shifter is held onto the tracks by four "clamps" while a "spud" is pushing down onto the ground with the result of lifting the track shifter and the track up into the air. The men on the track gang are using #2 shovels to level the ground beneath the ties. After the ground is leveled on both sides of the track the shifter will push the track outward to the edge of the dump so that the trains will be able to dump the waste rock over the edge. Note the much longer "tower tie" with the iron tower attached to it. At the top of the tower

an arm extends out to the center of the track where it holds the 750-volt wire that powers the locomotives. Joe Dunn touched his wrench to one of these towers which the 750-volt wire was also accidentally touching. Joe, a big man, was knocked to the ground and dazed; he figured it would have killed a smaller man (see "Saved From Electrocution"). Working on the track gang was almost a rite-of-passage for boys at Bingham High.

PEOPLE OF COPPERFIELD

There was a wide diversity of people in Copperfield. It is said that there were 27 different nationalities represented. Just the names of the different neighborhoods such as Jap Camp and Greek Camp are indicative of this diversity. Every group of people made positive contributions to society in Bingham. Their different habits, ways of life and foods added to the variety and interest of life in Copperfield.

Mrs. Gonzales will always be remembered for the terrific tortillas and beans that she made. It was surprising to me to see her frying tortillas right on top of the stove rather than in a frying pan. But the tortillas and beans tasted great. One of the ingredients was prickly pear cactus. Her family would go out into the hills and gather gunnysacks full of cactus that she included in their diet.

The Japanese kept mostly to themselves in Jap Camp but one memorable occasion was a PTA fund raiser at the Grade School. Several Japanese girls gave a dance demonstration at which they wore beautiful kimonos. All the people from Jap Camp came to see it and everyone was very well impressed.

Many of the Greeks were bachelors or at least led frugal bachelor-style lives here and sent money to their families in the old country. Many of the businesses in Bingham were owned by people of Greek ancestry.

Most of the other nationalities sort of melted together and we kids didn't know or care where they came from. The European countries were almost all represented but the only way you could tell was by their last names or by finding out by accident.

The real value of the people in Copperfield was that we mixed together and got to know each other and learned to appreciate each other. The kids all went to school and played together; there were no strangers. Most of the men worked together and often drank together. Point to any house in town and almost everybody could tell you who lived there. We bought groceries in the same stores, walked the same paths to and from our activities, and we were dependent upon one another for entertainment. Most people responded positively to this enforced camaraderie – especially the families with children.

Some people, however, weren't totally assimilated into full community life but had smaller groups of friends – this was especially so with the childless men in Greek Camp and Jap Camp. Very few women worked outside the home so those who weren't socially inclined could be left out of community fraternization; but they were still very good people.

Once, when a tourist who I was showing around had immediate need of a restroom, she went to the home of a Mexican family who welcomed her very courteously. The lady tourist was greatly impressed by the graciousness of Jose and Carlos's mother and how absolutely clean her home was.

The old lifestyle that we enjoyed in Copperfield has been replaced by a modern lifestyle in which a family can be much more independent and anonymous. A person can climb into his car, drive twenty miles to a big shopping mall, do his shopping, and return home without ever saying "Hello" to his next door neighbor. Then he can stay in his house, using his telephone for necessary conversations, and watch television, still without having spoken to his neighbor.

BINGHAM-MIDVALE ROAD DISCUSSED

Enthusiastic Meeting Upon This Subject By the Local Commercial Club
Good Speeches and Much Enthusiasm

An enthusiastic meeting was held in the rooms of the Bingham Commercial Club last Friday evening for the purpose of formulating plans for a campaign calling for the construction of a concrete highway connecting this city with the present paved highway terminating at Midvale.

The meeting was not attended by as large a crowd as would otherwise have been present were it not for the fact of two or three other functions of a public nature which were called for the same evening. Those present, however, were extremely enthusiastic upon the subject at issue and unanimously pledged themselves to its support.

The meeting was opened by Mayor Kelly, who stated the object of the gathering and introduced Mr. Wm. F. Long of Salt Lake City. Mr. Long gave a most logical and masterful address upon the subject of permanent roads, stating facts and figures as to both cost and maintenance. He also called attention to the rapid increase of heavy traffic and the necessity of building roads of such a nature as would come up to both present and future requirement. He quoted many government statistics and called attention to the great factor that the modern power truck was proving as a source of relief to the over-burdened railroads. Mr. Long proved a thorough master of his subject and was listened to with marked attention by those present. At the close of his address he exhibited a number of stereopticon pictures of the latest and most approved types of concrete roads in Utah and practically every other state in the union.

Mr. Wm. M. Wilson was then introduced by Mayor Kelly. Mr. Wilson referred to the wonderful mining interests of Bingham and how essential they were proving towards the winning of the war. He showed the possibilities of Bingham from the standpoint of the tourist and declared that a road such as was being discussed, would be the means of bringing to the city thousands of automobilists who would be anxious to see the wonder of the greatest copper camp in the world. "I know it is your intention to some day have a road of this nature," said Mr. Wilson. "but hell is paved with good intentions, and good intentions alone do not accomplish much in the paving line on this mundane sphere. The only way to get what you want is to get busy now and go out after it. If the people of Bingham will just put the 'pep,' push and the ginger into their efforts to secure this road as they do into other personal and civic matters, there is no doubt in my mind as to their final success." Mr. Wilson also gave some interesting facts and figures regarding the development of the motor truck and the relationship of that vehicle to modern business.

A committee of three was appointed to draw up resolutions to be presented to the proper authorities, and the meeting adjourned with the feeling prevalent that a good start had been made in a good and most worthy cause.

The above article was printed in the May 24, 1918 edition of The Press-Bulletin, Bingham Canyon, Utah.

CHAPTER FOUR

COMMUNITY ACTIVITIES

There were only a few organized activities that included the whole Community of Copperfield and when I was growing up most of those were down in Bingham – such as Galena Days and the Fourth of July parades and races. After I started going to the High School in Copperton there were more activities such as football, baseball, and basketball games that we watched. As for playing these sports, however, if the practices required an extra bus trip then Copperfield was often left out with the explanation that there weren't enough players to justify the expense of the extra bus run. The result was that it was impractical for students from Copperfield to participate in High School baseball, basketball or several other extracurricular activities such as Yearbook Editing.

The Copperfield Boy Scout Troop 112 was the one organized activity that most of us boys participated in as we were growing up. A few very dedicated men acted as Boy Scoutmasters and they were a tremendous help to all of us boys. Other than the Boy Scouts we made do with what was at hand. Saturday matinee movies at the Princess Theater down in Bingham were a weekly event; we sold ore to the tourists during the summertime and went sleigh riding in the winter. I personally spent considerable time hunting or just hiking through the hills but not many other kids or adults did this so it couldn't be classified as a Community Activity.

BOY SCOUT TROOP 112

There were six Scout troops in the Bingham District. They were Troops 111, 112, 113, 114, 136 and 150 (350 was also in Copperton for a while). Troop 111 was in Bingham (sponsored by and met at the Methodist Church in Carr Fork), Troop 112 was in Copperfield (sponsored by the PTA and met at the Grade School), Troop 113 was in Highland Boy (sponsored by the Methodist Church, met at Highland Boy Community House), Troop 114 was in Lark (sponsored by and met at the LDS Church), Troop 136 was also in Bingham (sponsored by and met at the LDS Church) and Troop 150 was in Copperton (sponsored by and met at the LDS Church -- was Troop 350 for a while). The Utah Copper Company (later Kennecott) unofficially fostered these troops; several of their management personnel, such as Vern Pett and Kay Landenberger participated in planning the District activities..

An interesting note about Troop 111 was that Tak Mochizuki, while in the Army in Italy during World War II, met his former Scout leader, Meredith Smith, who was a Minister of the Methodist Church in Bingham and was serving as a chaplain in Italy.

Troop 112 met once a week in the auditorium of the schoolhouse. Joe Dillier and Don Evans were the Scoutmasters for about a year each before they moved away. Malcolm "Scotty" Robertson took over as a steady, long-term Scoutmaster from about 1941 to 1948. Scotty dedicated seven years of his life to the weekly meetings, summer camps and other activities of Troop 112 and then continued to work as the Chairman of the Scout Committee until he moved down to the valley in 1958 where he again worked with the Boy Scouts. He was concerned about the boys and was a positive influence on the twenty-five or so boys who were in Troop 112 each year. With a turnover of about five boys per year there were roughly fifty-five boys in the troop during Scotty's tenure. Other Scoutmasters who followed Scotty included Nelson "Buck" Leyba for one or two years, Salvador "Chinky" Aguayo for three or four years and Lee Halverson who assisted Chinky for about one year and was himself the Scoutmaster for another year. Chinky and Lee kept the troop going even after they had moved from Copperfield by driving up from the valley. Chinky also became a Scoutmaster in Kearns and in West Jordan – he was in scouting for 30 years; during that time he took his Scouts to Yellowstone seven times and on three 50-mile hikes. Lee was also the District Chairman down in the valley for several years.

When Scotty first became the Scoutmaster of Troop 112 he worked for the U.S. Company and lived in Telegraph. He wanted to work for the Utah Copper but the hiring of the other's men by either company was frowned upon. Due to the involvement of Utah Copper management in the Boy Scouts Scotty got to know them and was hired by the Utah Copper and moved into a company house in Copperfield.

The Scouts really did work on advancement during Scout meetings and several of them reached the rank of Eagle Scout. Years later after everybody had moved down to the valley Chinky Aguayo and Lee Halverson, who had both attained the rank of Silver Beaver, were honored to participate in the Court of Honor for Scotty Robertson's youngest son, Kevin, when he received his Eagle award.

During Scout meetings the boys also played some rough-house games such as "Steal-the-Bacon", "Horses" and "Leap Frog". During World War II when scrap metal was needed for the war effort Troop 112 collected more scrap metal than any other troop in the Bingham District. They also collected newspapers.

The most popular of the other activities for Troop 112 was swimming. In the earlier years they went to Beck's Hot Springs or to the Municipal pool on the north side of Salt Lake and in later years to Saratoga Hot Springs down toward Lehi. They went swimming about six times a year. Everybody would load into the old 1929 White school bus that the Bingham District had purchased and drive to Beck's or Saratoga. The boys spent most of their time swimming, diving and splashing around for two or three hours in a disorganized free-for-all. But there was some actual swimming instruction and practice that took place. Most of the boys learned to swim at Beck's or Saratoga and they often held impromptu races and diving competitions. The boys were pretty well famished after all their exertions at the swimming pool so on the way home they would stop at a café in Murray, Midvale or Bluffdale for a hamburger and milk shake.

Garland had just received his Life Saving merit badge a few weeks before one of the swimming trips during which Paul Valdez was discovered at the bottom of the outdoor pool at Saratoga – he'd been there for quite a while and wasn't moving. Gar dove in and pulled Paul up to the bank. The boys lifted him out and laid him over the edge of the pool. They were preparing to resuscitate Paul when he came out of it OK without any further effort on their part.

There were some feelings between the troops. Copperfield boys were looked upon as socially inferior ruffians from "the other side of the tracks" while the Copperton boys were perceived by some to be goody-goody sissies. This difference in social status was brought home one summer when the troops went to Scout camp. The Copperfield troop, upon requesting use of the bus, was told that it was already reserved for one of the other troops. So what did Troop 112 use for transportation? The garbage truck!

Of course the truck had been washed quite thoroughly but it still had a certain odor about it and – it was still the garbage truck. The boys couldn't help but feel that they had been slighted.

Figure 108 – Bingham District Boy Scouts Swimming Meet Award 1943-44-45
Photo Courtesy of Kevin and Malcolm Robertson – This is the award for three years of swimming meets. Troop 112 won first place at all the swimming meets for about eleven years. Some of the woodburning on this plaque may not be legible so it is reproduced below.

TRACY WIGWAM

Bingham District Annual Swimming Meet
BOY SCOUTS OF AMERICA

TROOPS

111 112 113 136 150 350

1943

1ST

Troop 112

Garland Bray
Marion Carter
Peto Gomez
Gene Halverson
Lee Halverson
Rex Leatherwood
Teddy Leatherwood
Nelson Leyba
Billy McIvor, Capt.

2ND

TROOP 150

3RD

TROOP 136

1944

1ST

Troop 112

Salvador Aguayo
Garland Bray
Ronald Burke
Fino Contrares
David Gonzales
Lee Halverson
Gene Halverson
Max Ivie
George Kubota
Tom Korologos
Pete Leyba
Leon Miller
Billy Nevers
Tom Sotiriou
Gus Sotiriou
Paul Valdez
Rex Leatherwood
Captain

1945

1ST

Troop 112

Garland Bray
Ronald Burke
David Gonzales
Lee Halverson
Pete Leyba
Gus Sotiriou
Tom Sotiriou
Gregory Tyson
Paul Valdez
Rex Leatherwood
Captain

2ND

TROOP 111

3RD

TROOP 136

A great rivalry between the troops was apparent at the annual swimming meet. Each summer there was a swimming meet at Tracy Wigwam in Mill Creek Canyon. The swimming pool at Tracy was 20 yards (60 feet) long and the water was really cold – it wasn't heated at all. Right from the beginning Troop 112 had some exceptional swimmers such as Billy McIver, Rex Leatherwood, Duggie Gonzales, Lee Halverson, Paul Valdez and Garland Bray. Year after year as one of Troop 112's best swimmers left the troop another "unknown" would rise to the occasion. One year Troop 112's best swimmer, Rex Leatherwood, was sick. Scotty took Lee aside and told him to swim the whole length of the pool without breathing – just keep his head down and swim for all he was worth. Lee did as instructed. He kept his head down and swam as hard as he could. He couldn't see how the other swimmers were doing but when he reached the far end of the pool he had won first place!

Year after year the same troop won – Troop 112! This was very galling to some of the other troops and to their Scoutmasters in particular. So they conspired to introduce "special events" with which they hoped to change the outcome of the swimming meet. One of the troops thought they had some good underwater swimmers who could beat Copperfield's best so the next swimming meet included underwater swimming for the first time. The opposing troop's best was able to swim the entire 60-foot length of the pool underwater. This was farther than any of the preceding swimmers had been able to do. But then it was Paul Valdez's turn. Everybody watched, almost holding their own breaths, as Paul, deep under the water, approached the far end of the pool – touched it, turned, and swam halfway back to the other end! He had beaten all previous swimmers including the best from the other team. But then, as if to rub salt in the wound, Lee Halverson took his turn. Lee swam the entire length of the pool, turned, and swam all the way back to the far end! Lee and Paul were first and second in the underwater swimming competition. The strategy of introducing this new event had backfired on the troop that had proposed it. Copperfield had beaten the other troops more handily than ever! Another attempt at introducing a new event was Diving. Again it backfired – Garland took first place in the Diving contest.

The next attempt to find a way to beat Copperfield was to change the rules so that the adults from each troop could participate along with the boys. A meeting was convened at which this change was considered. During the meeting, however, it became apparent that Lee, who was the Assistant Scoutmaster and Garland who was swimming for the University of Utah, could both participate for Troop 112 if the rule was changed. These two could both beat anybody that the other troops had. The proposal to change the rules was immediately dropped. Lee and Gar gave an exhibition swim rather than racing with anybody.

Thinking back, years later, about why Copperfield always won the swimming meets there could be a couple of explanations. Copperfield was at an elevation of about 6800 feet up in the canyon where the streets and trails were steep and almost all walking involved a climb. Copperton was 1300 feet lower at an elevation of 5460 and the streets and trails were all relatively level. The boys in Copperfield probably had greater lung capacities and were in better shape because of all the hill-climbing that they did. Another explanation might be that the Copperfield boys simply tried harder.

Boy Scouts quite often do some funny things. Following are some of the "screwball" things that I remember happening in Troop 112. One year the troops from Bingham went to camp for a week. All of the groceries were purchased through Leonard Miller's store and a menu was provided for each meal. Each troop had its own camp site, its own portion of the groceries and did its own cooking. There was some room for modifying the menu. When black bean soup was called for early in the week it didn't sound very appetizing so they found a substitute for it. A day or two later black bean soup was again called for and again they substituted for it. This was repeated several times. The last day at camp was the day of reckoning. There was almost nothing left to eat except black bean soup! They managed to choke it down but only because they were really hungry and had no other choice.

A few of the kids who were in Troop 112 led pretty tough lives at home. This became apparent when they went on a one-week camp to Tracy Wigwam. They all loaded into the old bus that the District had purchased and away they went to camp. It was a regular, fairly normal camp with the boys taking turns at the cooking and the food wasn't really outstanding. The boys got along pretty well with each other about the way you would expect them to. At least that is the way most of the boys felt about it. But for one boy this was Heaven compared to his life at home. When it came time to go home he didn't want to go! His solution was to take an axe and chop a hole in the gas tank of the bus! This created a delay but the boys were all delivered safely back to Copperfield.

Most of the boys had some memorable personal experiences at Tracy Wigwam. Competitions in the swimming meet were remembered by all with many of the boys becoming champions and enjoying victories they would never forget. My own major victory came when I won the backstroke race. An older boy had taught me the backstroke only a week or two earlier. This older boy was in the race with me and I beat him along with all the other competitors!

Doctor Richards cooked hamburgers for everybody after the swim meet. These always tasted great to the hungry boys but they were also remembered as being half raw. Watermelon was the other main dish at these meals. But there were some other odd experiences that would stick in a boy's mind for the remainder of his life. One of Garland's experiences was his first and only encounter with a rattlesnake. Gar was walking along a dirt trail close to camp when he came upon a medium-sized rattlesnake that was in the trail. He warily walked around the rattler and hurried back to camp with neither he nor the snake bothering the other. Gene Halverson also had experience with rattlesnakes at Tracy Wigwam. He killed two large rattlers in one day just below Tracy.

My odd experiences at Tracy include the doctor at camp determining that I had athlete's foot and couldn't go swimming until it was cleared up. Not to worry though – the doctor had me soak my feet in a bucket of water with a small amount of Clorox. The athlete's foot was cleared up in only a day. When I hear of the high-priced fancy treatments for athlete's foot now days I still think back to the Clorox cure and how effective and inexpensive it is.

Another humorous experience was when the troop, while at Tracy Wigwam, went on an early-morning hike up the canyon. The sun was just coming up as we hiked up the dirt road and came upon an unexpected sight – a guy and his girl parked in a car. They obviously had been there most of the night in what they considered to be a private "Lover's Lane". Now, to their dismay, they were confronted by the sight of a whole troop of Boy Scouts marching upon them. There was a lot of scrambling inside the car and then the engine came to life. As the whole troop of Scouts gawked in curiosity and astonishment the car with the guy and his girl roared away amid a big cloud of dust. I've sometimes wondered whether that poor girl was embarrassed for the rest of her life or whether she got over it in later years.

Camp Steiner was very different from Tracy Wigwam. Steiner was in the Uinta Mountains at an elevation of about 10,400 feet. There were numerous lakes and streams that were loaded with trout. Each troop had its own open-front log cabin with bunk beds. There were Scouting activities in which the boys could earn merit badges but fishing was the main activity for many boys.

I especially enjoyed the fishing but it did lead to a mishap. When boarding the bus for Camp Steiner one year I wore a hat to which I had attached several fishing flies. My friend Warren was pushed up against me from behind by the jostling boys and was "hooked" on his eyelid by one of my flies. The doctor used some wire cutters to cut the barb off from the hook and then he withdrew the body of the hook with no problem. It was a close call though and could have been very serious if the hook had pierced his eyeball just an eighth of an inch away.

THE CRACKER BOX

By Salvador (Chinky) Aguayo

The old bus had served its purpose many times over. Its service as a school bus was years behind it. Somehow its life had been prolonged by the Bingham Boy Scout District which had purchased the bus from the School District to use in a variety of the activities of the five different troops in the District. Each troop could reserve the bus for any activity that they wanted. The bus was put in the loving hands of Joe Timothy, the District Chairman, who also drove a school bus for the High School during the week. So Joe was in charge of maintenance also. The bus was known by everyone as "The Cracker Box" and the Scouts looked on it much like they would look at a venerable old grandfather. They appreciated all that the bus did for them. They knew that most of the fun places they were able to go to would not be possible without "The Cracker Box". Everyone was always ready to push the bus when needed. Sometimes the bus needed to be pushed for miles and people would come out of their homes to watch twenty-five kids pushing a dilapidated old bus down the road. The boys never seemed to mind having to push the old bus; after all, it was just our turn to help an old friend.

When I was older and became a Scoutmaster of Troop 112 in the Bingham District, I would take my turn in reserving "The Cracker Box" for many of our activities including camping and swimming. I include swimming because there was no place to swim in Bingham. The only swimming pool was in the Old High School and it had been condemned by the Health Department. We had a choice of two public swimming pools we could go to if we had the transportation. One was at Wasatch Springs in Salt Lake and the other was in the oppose direction in Lehi (Saratoga). Most of the time we preferred to go to Wasatch Springs because we could get drinks or a snack on the way home. In those days there were no fast food restaurants as we have today but they had a few dairy stops where you could get milk, orange juice, chocolate milk, or even a soda pop was sometimes obtainable. By stopping at a drugstore in Murray or Midvale you could get a sandwich. It was always an adventure to take a bus full of Scouts anywhere because the outcome of the trip was always in doubt so we always made sure we had the telephone number of someone who could organize a rescue party.

An unusual chore we gave "The Cracker Box" once was a time when the Troop wanted to have a paper drive so for several weeks we collected piles of paper and we reserved the bus to haul the paper to the recycler in Salt Lake. The bus was crammed full with paper so I could only take two scouts along to help. I thought we should have enough to buy a new bus but we were lucky to pay for the gas it took to get there and back. I decided right there never to collect paper again.

SATURDAY MATINEES AT THE PRINCESS THEATER

I can barely remember the Dianna Theater in Copperfield; it was shut down in about 1939. The Princess Theater in Bingham, on Main Street just below Carr Fork, was the showhouse of my growing-up years. Every Saturday they would have a serial matinee. We kids would walk or bum a ride through the tunnel to Bingham, go to the show and then either walk or bum a ride back. The movies were generally serials that would end at a particularly exciting point, to be continued next week and the next over a period of several months. Tarzan, The Lone Ranger, Zorro, Batman and Robin, and Superman were some of our heroes. They also had serial detective or mystery movies that you could win a free ticket to if you correctly guessed who the killer was.

SELLING ORE TO THE TOURISTS

The Utah Copper open-pit mine was a major tourist attraction during the summer months. Automobiles from all states in the union would drive up the streets of Bingham, through the 1 ¼ mile long vehicular tunnel with its 6% grade and emerge in Copperfield to see the mine. The Greyline Bus Tours also brought many tourists in their buses to see the mine. The mining company had an observation point for visitors in the canyon bottom on the edge of the mine. Sometimes there was a representative of Utah Copper who used a loudspeaker to tell the tourists about the operation; he knew all the statistics of the mine – how far it was from top to bottom, how many tons of ore and waste were mined per day, grade of the ore, etc. We kids picked up all the info that he provided to the tourists and some more besides.

We kids offered two services to the tourists. Those who drove themselves to the mine were prospects for our guide service. Otherwise they, along with those who came in the buses to the observation point, might want to buy some ore specimens as souvenirs. During the summer and especially on Saturdays and Sundays it could be a thriving business for ten or more of us kids. We would take turns approaching the tourists' cars as they emerged from the tunnel and would offer to guide them to the open-pit copper mine. Once the tourists arrived at the observation point, whether by bus or cars, they were fair game for all of us at once to sell ore specimens to.

If a carload of tourists wanted a guide then we would get in with them and direct them to a vantage point where they could get a much better view of the mine. We would then go into our spiel regarding the history and statistics of the mine. And we also would offer them some ore samples. There were several vantage points of choice – one of my favorites was on a mine level way up on the West Side of the mine. These levels were only actively mined by a contractor for a few months at intervals of about two years and were essentially deserted in between contracts.

The spiel regarding the mine statistics went something like this: The first mining claim in Utah was filed here in Bingham Canyon in 1863 on silver/lead ore in Galena Gulch. The open-pit copper mine was started in 1906 by Daniel C. Jackling. It is 1890 feet from the top to the bottom and is nearly a mile across. Production from the mine is 60,000 tons per day of ore (it increased over the years). They have to strip two tons of waste from the upper levels for each ton of ore that is mined. The grade of ore is 0.8% copper. They also recover molybdenum, gold, and silver. There are thirty-two electric shovels with six-yard dippers. The ore cars hold 90 tons. The ore is hauled by rail to Magna-Garfield for milling and smelting. The waste cars hold 40 tons each. The waste is hauled out to the waste dumps farther down the canyon. As the mine goes deeper it is expanded outward so that more of the town is mined out each year. When the main road up the canyon was dug out they made a tunnel to connect the main part of Bingham with Copperfield. The tunnel was dug in 1938. It is a mile and a quarter long with a 6% grade. Our spiel was considerably longer and included pointing out various features of the operation. After viewing the mine we would return to the upper portal of the tunnel and the tourists would give us a tip. The tips varied from about 25 cents to as much as a dollar. That was a good income for us kids!

The sale of ore specimens was also a lucrative business. We could identify peacock copper, galena lead, iron pyrite, blue vitriol, quartz and other minerals. To demonstrate that blue vitriol contained copper we would spit on a knife blade and rub it on the blue vitriol so that the iron was replaced by copper. I'm sure this impressed them either by our knowledge or by our uncouth manners.

"FINDING" ORE TO SELL

We kids got our ore samples from several sources. Sometimes our Dads (especially the powder monkeys) would bring ore specimens home from the mine or we would occasionally go into the mine ourselves and dig them out. Mostly, however, we got our ore samples from the ore cars of the U.S. underground mine.

The periphery of the Utah Copper open-pit mine was underlain by hundreds of miles of underground workings from which lead, zinc, silver, gold and copper were mined. The U.S. mine was one of the largest. The portal to the Niagara Tunnel of the U.S. Mine was in Copperfield near the upper portal of the vehicular tunnel. The ore from the U.S. Mine was hauled in small rail cars out of the mine and onto a trestle beneath which were larger, standard gauge railroad cars that would be hauled to the mill down in the valley at Midvale. Ore from the small rail cars was dumped from the trestle down into the larger cars. These larger cars would sit beneath the trestle for several days before they were all loaded and hauled away. We kids would sneak up into these ore cars and search through the broken ore for particularly good-looking samples. It was like treasure hunting! Galena, pyrite, and quartz crystals were the most common minerals to be found there. Sometimes we took a hammer with us and broke the prettier parts off the larger rocks. Other times we raised the larger rocks above our heads and smashed them down onto the railroad track to break them up. We had to be careful not to be in the ore cars when they dumped ore down from the trestle above. I never heard of anybody being caught or injured in this way. We didn't have any real problems with watchmen either. Probably the U.S. Company knew that we kids took ore samples from their cars. Either they were well-disposed toward us and so looked the other way or they figured it was less of an expense to have us swipe a tiny amount of their ore rather than pay for a watchman.

An ore-gathering expedition that I especially remember was when three or four of us went into the open-pit mine to where some old underground workings had been exposed. This was on the East Side in the Ohio Copper area. The old abandoned workings had been squeezed and caved in over the years so that there was only a small opening about a foot in diameter that showed on the face of the level. Around the opening and in the debris pile below the opening there was ample evidence of blue vitriol. We took turns getting samples of the ore mineral. It was a matter of shoving our arm into the hole and groping around until a stalactite was located and then breaking it off. Each of us collected several good samples of the pretty blue ore. In later years I learned that blue vitriol was actually copper sulfate or the mineral chalcanthite and that good, solid pieces would have formed in the absence of air. The old, caved-in underground mine was a perfect place for copper sulfate to crystallize. The solutions rich in copper slowly moved downward through the rock and there was no air.

THE COPPER WATER

The Copper Water was known by most of the kids in Copperfield as an almost magical place with green water where they could copper-plate nails, knife blades and tin cans. It was located in the tailtrack of Fox's Switch by the upper end of Main Street. The water issued from an old mine about halfway up the 40-foot rock face at the very end of the tailtrack. It formed a small creek that flowed along the level for 40 yards or so to where it ran through a wooden grate and was piped underground – probably to a copper precipitation plant near Lead Mine. There were a few small pools in the creek that were handy for immersing larger metal objects like big nuts and bolts and if they were left for several days they would be completely replaced with copper.

When the open-cut mines of the Boston Consolidated and the Utah Copper companies were started in 1906 they hauled their waste rock to the south and filled in the upper portions of the gulches all the way around to Galena Gulch (the U.S.) which was the main branch of Bingham Canyon. Galena Gulch was partially filled in so that the north side was completely covered, the bottom was buried to a depth of a few hundred feet and the dump extended several hundred feet up the south side. The mouth of Galena Gulch, prior to the mine dump, was about half a mile above Fox's Switch. Copper that was leached out of the waste rock by natural precipitation drained to the old creek bed of Galena Gulch and then down-canyon toward Fox's Switch. The outlet at Fox's Switch was actually a very old mine working along the bottom of the creek bed that had been covered by the waste rock and soil; some pieces of timber could be seen at the outlet.

The copper-coated nails that the kids produced served mostly to impress tourists but some of them might have been sold as souvenirs. The kids also impressed tourists with their ability to coat a knife blade with copper by spitting on the blade and then rubbing it with copper sulfate (blue vitriol). It wasn't always clear whether the tourists were more favorably impressed by the production of the copper or negatively impressed by the uncouth act of the boy spitting on the knife blade.

The economic possibilities regarding the Copper Water were recognized very early in the copper mining history of Bingham Canyon. Several precipitation plants were operated during the decades since the first underground mines penetrated the Copper Hill. The advent of open-cut mining resulted in low-grade copper-bearing waste being dumped into gulches where the copper in the broken rock was subject to leaching by rain and snowmelt. The resulting copper-laden water drained down to the buried stream beds and ran along the channel until it came to the surface in places such as the Copper Water in Copperfield. As the waste rock from the open-cut mines grew in volume the waste dumps became bigger and extended farther along the mountain sides into more gulches. The number of sources of copper water running into the Bingham Canyon drainage and the amount of water issuing from them steadily increased over the years.

Figure 109 – The Copper Water in Fox's Switch Tailtrack
Photo Courtesy of Shirl and Isabell Scroggin – An opening and some old mine timbers can be seen at the place where the water issues forth. The water is green. There also is a yellow and green encrustation of copper and iron mineral salts on all the ground that can be seen in the photo – probably from the spray or mist caused by the splashing of the mineral-laden water. The original purpose of the old mine adit along the bedrock-gravel interface isn't known but it could have been for mining placer gold.

Precipitation of copper from the mineral-laden water was a process that yielded a good profit. So the amount of water from natural rain and snow was supplemented by the addition of water from other sources to the upper surfaces of the waste dumps and also by the addition of sulfuric acid which hastened the leaching of copper from the waste rock. The production of copper from leach waters continues to be a significant source of income to the Kennecott Utah Copper operations.

SLEIGH RIDING

Copperfield had some ideal conditions for sleigh riding -- but with hazards. The good thing was that the roadways were steep, a lot of snow fell during the winter, and they plowed the roads (this made a good, compacted base that was very slippery). The bad thing was that cars and trucks used the same road as the sleighs. Some sections of the road were steeper than others and some sections had more vehicle traffic than others. If you didn't let the hazards stop you then it was possible to ride a sleigh down the road for over a mile without stopping. The limit on how far up the canyon a kid could start his sleigh ride

Figure 110 – Snow by U.S. Hotel In 1952
Photo Courtesy of Eugene H. Halverson - The snow apparently fell during the previous day after which the road was plowed while some cars were parked against the curb. Additional cars then parked wherever they could find a spacc. Kids are going down the street after school. The plowed road was perfect for sleigh riding. The vehicular tunnel is to the right, off the photo. Judging from the automobiles and remembrance of a heavy snow year this photo was taken during the winter of 1952-53.

Figure 111 – Unplowed Snow on Main Street In 1952
Photo Courtesy of Eugene H. Halverson - One lane has been plowed but considerable snow is still on the roadway. This was an especially heavy snowfall. Snow on the road is rather loose and not as hardpacked or slippery as it usually was after being plowed; maybe some sand was mixed into it. This loose snow would have been "OK" for sleigh riding but not nearly as good as hardpacked snow.

was, at least for me, the overly steep grade between the old U.S. and upper Telegraph. This stretch of road was steeper than any other and there was a sharp left-hand curve at the bottom of the steepest grade just as it entered Telegraph. I was afraid of going too fast and not making the curve.

A perfect ride would begin at the top of Telegraph just below the sharp curve. It was best to have a long enough sleigh for your whole body to lay on with your boots sticking past the end just in case you had to dig your toes into the snow to slow down or turn. The ride could start by taking a few quick steps, then slamming the sleigh down on the snow and belly-flopping onto it an instant later. Another way was to lean over the back end of the sled with one hand on each side of it and run for several steps before belly-flopping onto the sled. You could get off to a quick start either way.

The first quarter mile of the ride was steep and fast. There were a couple of easy turns by Emil Bodmer's house that were easily negotiated. Then the road leveled off a bit at the lower row of houses in Telegraph so that the sleigh slowed down but still kept going about as fast as a kid could run. It was here that a dog started chasing me and I punched him in the nose with my fist. Immediately below

Figure 112 – Paul and Vivian Halverson on Sleigh About 1940
Photo Courtesy of Eugene H. Halverson - These young sleigh riders were quite happy and enthusiastic. They were probably being pulled along a near-level roadway. However, the really serious sledders would have presented a very different picture of determined older boys lying flat on the sleigh and racing down the roads at breakneck speed, alert for oncoming cars, sparks shooting up from the runners as they went over any spots where the road was dry.

Telegraph, however, at the Cocoa Dirt (old, iron-stained mill tailings) the road steepened considerably. The grade was so steep that most cars had to be in first gear as they came uphill. The sleigh really picked up speed starting here and for the next half mile. You might have to drag your toes to slow down enough to stay in control.

There was a narrow concrete railroad overpass (H-level of the mine) between the side roads to Dinkeyville and to Jap Camp with a curve just above it so that you couldn't see ahead of time whether there was anything in the roadway or not. I met an ore truck that was coming uphill in the overpass one time and there was just barely room between the wheels of the truck and the concrete of the overpass for me to slip through. I almost scraped my hand against the concrete as I zipped past the big wheels of the truck. There was no time to think – I just did it! Of course it didn't snow under the overpass but the wind did blow some snow in. Parts of it were bare roadway and would, with sparks flying, slow the sleigh down but not stop it. But immediately below the overpass the sleigh would rapidly pick up speed again. There were a couple of escape points from this stretch of steep road – above the overpass there was a road cutting to the right to Dinkeyville and below the overpass there was another road cutting to the right to Terrace Heights. However, most of the time we continued down the main road without turning off.

The road became less steep from Fox's Switch on down but there was a greater likelihood of encountering cars and the road was still steep enough that the sleighs kept moving at a high speed. And then came the U.S. Hotel and, across from it, the mouth of the tunnel. This was a good place to stop because there weren't any good stopping places from here on down to the end of the road. One way to stop was to go into the vehicular tunnel where there wasn't any snow and the runners of the sled would send sparks shooting out as the metal of the runners ground over the rocks in the hard surface of the road and we came to a halt. Very often a kid would be stationed at the mouth of the tunnel to watch for cars. He would flag the sleigh riders down if there were any cars about to come out of the tunnel. The usual way to stop if one really had to was to turn the sleigh sharply so that it would slide sideways with the runners digging into the snow but there was a risk of putting too much stress on the runners so that they would bend or buckle and ruin the sled.

The ride from Telegraph to the tunnel would take four or five minutes and then, to take another ride, you had to walk and pull the sled uphill for twenty or thirty minutes. The elevation of Copperfield was about 7200 feet with winter weather often so cold that you could only stay out for an hour or so – especially at night. The lobby of the U.S. Hotel was very often a haven where we could warm up.

This description of sleigh riding in Copperfield is only my version from a limited part of Bingham Canyon and only for a few of the seventy or so years that kids rode sleighs there. There were certainly many experiences by other people from other parts of town and in years before and after that have never been written. This is attested to by the scars on my father's leg. They were caused by him being impaled by the runners of sleds in two accidents that would be lost from memory if they weren't mentioned here. In one of these accidents cloth from his pants was carried into the deep puncture and caused infection. The doctor performed a gruesome operation to remove the cloth from the infected wound that was turning gangrenous. Other kids crashed into walls and automobiles or were run over by cars, sometimes with disastrous results. The sharp back end of the runners was a particular hazard because if a kid ran into an immovable object then the kid would keep moving forward after the sled had stopped and too often the runner would be thrust into his thigh. The other major hazard was that of the kid's head slamming into the car or wall or whatever the sleigh had hit.

My brother Garland had a paper route and two of his customers lived in Telegraph. Our dog, Tip, accompanied him on the paper route. After delivering all the papers in Copperfield (including Jap Camp, Greek Camp, Terrace Heights and Dinkeyville) then Gar would tie a rope from Tip's collar to the sleigh and have Tip pull the sleigh up to Telegraph. The last customer was Halversons up at the very top of Telegraph. After delivering this last paper then Gar would untie Tip and ride the sleigh back down the steep road to Terrace Heights where we lived while Tip raced along behind, running as fast as he could trying to keep up.

CHAPTER FIVE

NEIGHBORHOOD ADVENTURES

A real feeling or picture for life in Copperfield is made more complete by recounts of adventures of the inhabitants. The following stories are certainly only a small part of many exciting and interesting experiences; unfortunately most of them have already been lost as individuals have passed away.

BB GUN SHOOTOUT

It wasn't exactly like the OK Corral. Nobody got killed because they were using BB guns instead of .45s. They weren't aiming for the heart either – they were aiming at each other's ears. Marion and Bob faced each other from a distance of fifteen feet and took turns shooting at each other's ears. Bob stood up straight and bravely faced his foe as Marion carefully aimed. And then "Ping!". There was no outcry of pain as the BB hit Bob's ear and drew blood but he couldn't help flinching slightly. Then the roles were reversed and Bob had the opportunity for retribution. He aimed carefully for Marion's ear and "Ping!" he scored a direct hit and Marion's ear was bloodied for the third time. With both of them being right-handed, it was easier to aim at the other's left ear because it was more exposed. If they were to aim at the right ear there would be too much chance of hitting the other in the eye and they didn't want to do that. Each of them took three or four shots and made a hit almost every time. But neither of them had chickened out so they decided to call a truce.

This encounter took place on the road between the Upper and the Lower Terrace Heights on the far end above Bert Ivie's house. It wasn't clear just why they had started fighting but it was a very novel way of settling their differences. There were less than half a dozen other kids watching this exchange. There were no grownups. Both Marion and Bob were about thirteen years old at this time. In his adult years Marion worked in law enforcement; maybe this was the first practice session for his chosen career.

CARBIDE "BOMBS" IN GLASS BOTTLES

Carbide, when mixed with water, generates acetylene gas. The acetylene gas can then be burned in miner's lamps (this was a major use before the advent of electric lamps). Acetylene welding is another use for it. However, the gas can also explode if it is confined and this is what we kids used it for most often.

Leon Miller still carries scars on his left hand from a novel form of entertainment that he was involved in. Leon and four other boys went fishing with carbide down on the Jordan River! The idea was to set off underwater explosions that would stun or kill fishes so that the boys could catch them. They had a good supply of carbide and some glass bottles. The explosions would be most effective if they went off deep in the water. They put rocks into the bottles so they would sink, added some water and carbide and then screwed on the lids and threw them out into good-looking holes. The first few bottles sank pretty deep before they exploded and sure enough, some fish came floating to the surface. A couple of the kids were in the water so that they could catch the fish and a couple other kids were on the bank preparing and throwing the "bombs". Leon was in the water. They must not have put enough rocks into this one bottle because it bobbed back to the surface. Leon swam over to push it down deeper into the water. He grabbed the bottle, pushed it down and had just barely released it when it blew up. The glass cut his left hand quite severely and it bled badly until they got it wrapped up. Of course that ended the sport for the day and they took Leon home. He didn't go to a doctor or have any stitches but he still carries the scars.

Another incident was when the boys were blowing up bottles in a very small pond of water on H-Level. The water was only about five inches deep with four inches of mud on the bottom. The idea was to put some carbide into a glass bottle, dip about an inch of water into the bottle, screw the lid on real fast, and then push the bottle down into the mud so that it would stick there and then the boys would back off thirty feet or so to watch the explosion and the eruption of mud and water. This one time, however, the bottle floated back to the surface. So the boy pushed it back down again and again but the bottle exploded before he could get very far away and he was cut by the flying glass.

THE BOY WITH THE PAINTED HEAD

Every community should have a huge block letter of painted rocks on the mountain above their town to proclaim the town's identity. Such a letter would distinguish the community and be a source of pride for all to see. It should be visible from a mile away. Several of the boys decided that Copperfield needed such a letter. A "C" for Copperfield on the side of the mountain above Dinkeyville would be just the thing. All the people in town would wake up one morning and be astonished at the sight of a big "C" looming above them where previously there had been only a bare patch of ground on the hillside.

The boys would first cut and remove a small amount of grass or brush as necessary. Then they would gather the rocks from a large area around the chosen site and place them to form the huge "C". What they really needed, however, was enough paint or whitewash and some brushes with which to paint the rocks. And they knew where they could get them.

The expansion of the open-pit mine would soon necessitate digging out the lowest part of Copperfield where several businesses were located. The Independent Grocery was one of the businesses to be mined out and it was now vacant. Except that the crew that maintained the rental houses for the mining company was now storing their paint and other supplies in the otherwise empty building. There were 5-gallon buckets of paint of various colors, some full that had never been opened and others that had been partially used. The painters had put a padlock on the front door but there was an open window at the back of the store. The window was about thirty inches square and the bottom of it was six feet above the ground.

Two of the boys crawled up through the window and dropped into the store. They looked through the painting supplies and found a two-thirds full bucket of white paint. It was just what they needed so they carried it over to the window and hoisted it up to the window sill. A third boy waited outside to lower the bucket of paint to the ground. He saw the bucket of paint in the window and raised his arms up to receive it. One of the boys inside the store pushed the bucket farther out but the outside boy didn't have his hands on it yet. The bucket tipped outward and the loose lid came off. The outside boy, who was gawking upward, was helpless to stop the deluge of white paint as it cascaded down onto his head and face.

The next day and for a week after that the "outside" boy had white paint in his hair and was tight-lipped about what had happened. Needless to say, the boys gave up their project. Copperfield was slowly chewed away by the open pit mine until the town was completely gone without ever having a "C" on the mountain above it.

SHOWER WHERE YOU CAN

With Lee Halverson

Modern conveniences were a hit-and-miss thing in Copperfield. Some of the older sections of town had been built without running water in the homes so that the inhabitants had to carry all their water in buckets. The lack of running water also made outdoor crannies the rule in these areas. Other houses were mid-way between the old-fashioned and modern -- some had running water but no flush toilets while others had flush toilets but no bathtubs or showers. The more fortunate families had all the modern conveniences of running water, flush toilets, bathtubs and showers. The **really** upscale homes (and there were very few, mostly in the lower part of town) even had lawns.

Most of Jap Camp and all of Greek Camp didn't have running water or the accompanying conveniences in the individual apartments but each camp had a shower house that was primarily for the male workers. There were a few families in Jap Camp who also used the shower house but only during certain hours. The men who lived in the individual apartments in Jap Camp and Greek Camp, since they didn't have running water, used wash basins for simple hand and face washing but went to the shower house for all their other needs. Dinkeyville was almost completely devoid of showers or bathtubs. Telegraph had a mixture – some houses had no running water at all and others had all of the conveniences including bathtubs or showers. The men who lived in houses that lacked showers or bathtubs, as in Dinkeyville and Telegraph, generally showered at work – the U.S. workers at the U.S. Mine, the Utah Copper employees at the facility in Greek Camp. Families who lived in houses that didn't have bathtubs or showers could always heat water on their stove and take a bath in a washtub in their kitchen. Some of them had to carry their water in buckets for quite a distance so it was a real chore to take a bath. It was a relief when the kids, both boys and girls, went to school down in Copperton where they would shower as part of their physical education classes. The women and young girls were generally stuck with the washtub in the kitchen routine. Another alternative for everybody was to go to a friend's house and use their shower. This was generally safe and was convenient if their friend lived nearby.

The boys who wanted to take a shower were quite enterprising and opportunistic. They went wherever they could to take a shower. Some of the boys went to the shower house in Jap Camp but this could be embarrassing because a woman might come in. The shower house in Jap Camp was supposed to be only for the residents there – they called these venturesome boys "strays". Greek Camp was a safer place for boys to shower because there were no women there. When the boys went to Greek Camp it had to be when the men weren't going to or coming from work – they generally went to take a shower in the middle of the day when only the watchman was there. The Gemmell Club down in Carr Fork also had showers so boys would often go there to play basketball or other sport and then shower afterward.

There was another possibility that Lee Halverson from Telegraph was able to take advantage of for a few years. The dry house at the Silver Shield mine had hot showers for the miners. Lee would visit his friend up at Silver Shield and take a shower while he was there. There was an ample supply of hot water so there was plenty of it for him to delight in. He could stand under the shower of hot water as long as he wished. But there was one problem. The water tasted terrible because it contained arsenic! The water for the shower came from the mine drainage and probably had a lot of metals in it besides arsenic. Anyhow, while showering, if he let any of the water into his mouth the taste was terrible. This water wasn't used for drinking by the miners or in any of the homes. The drinking water came from another source.

While Lee and his buddy were hiking around on the hill above the Silver Shield mine they became thirsty. Several small springs on the hill fed a small stream that really looked enticing. The water was clear and cold and it was a long ways to where good drinking water was available. So they drank some of it. For the first second or two the water tasted sweet but then a bitter metallic taste like arsenic became apparent. But this still didn't stop them from trying to drink that water. Lee's brother Gene told him of a way to take care of the bad taste – suck it through a straw made from a branch of an elderberry tree. Lee tried that – he cut a ten-inch joint from the branch of an elderberry tree. The inner part of an elderberry branch is a soft, cream-colored pithy material and there is usually a small hole in the center that runs between joints. He pushed a nail and a wire through the center to clean out the hole and to make sure it was continuous. The material in the center was supposed to remove all the arsenic. He sucked the water through this homemade filtering straw and – it didn't work. The water tasted as bad as ever.

Fortunately, Lee is still alive today. The arsenic wouldn't be a health hazard unless a person drank a significant amount of it. But some of the people who showered in the arsenic-laden water didn't know it wouldn't hurt them and it worried them for years afterward. Lee was given another concern when Gene told him several years later that elderberry plants, including the wood, are high in cyanide.

A DEER IN THE BACK SEAT
With Eugene H. Halverson

Gene was driving his Dad's 1936 Ford between Copperton and Bingham one night. Garland was in the front passenger seat and Nuffy Korologas was alone in the back. There was another car in front of them and as they approached the English Dairy at Dry Fork the car in front hit a deer. The deer lay alongside the road as if it was dead.

Gene stopped to pick up this free deer meat that was as fresh as could be. It was a yearling doe and would be real good eating. He picked up the deer and tried to put it in the trunk behind the back window but the trunk was only a foot wide and the deer wouldn't fit. So he threw it in the back seat with Nuffy and prepared to drive away. But the deer "came to" and was likely to start thrashing around in the back seat.

They decided that Nuffy would sit on it and hold it still while Gene drove up the canyon. Nuffy was able to control the deer -- at least for a while.

Just as they got to the Princess Theater in Bingham Gene had to stop for the traffic and people came streaming out of the show house. Then the deer got loose and started jumping around and flailing its hooves at which Nuffy started screaming. The deer jumped back and forth between the back and the front seats. Gene and Gar tried to hold the deer as it thrashed around. The people coming out of the theater were trying to look into the car to see what the screaming and wild activity was all about. Finally the traffic cleared and Gene drove a little farther up the canyon away from the theater. Between the three of them they got the deer back under control with Nuffy sitting on it again. Gene drove through the tunnel and all the way up to Telegraph where he lived.

After getting home with the deer Gene dispatched it and took care of the carcass like any good hunter would. He hung the fresh meat in the mine tunnel that was by their house where it was nice and cool. What did his dad say about this event and the fresh meat? He said "Thank you".

UP THE SHAFT AFTER SCHOOL
With Lorraine Clinton Beckstrand

The Copperfield School was close enough to where almost all the kids lived so they walked to and from school. The exceptions were the half dozen kids who lived way up in the Old U.S.. It was over three miles from the school to the U.S.. The kids were driven to and from school in Hy Johnson's

Figure 113 – Students From U.S. With "Bus" and Driver Hy Johnson
Photo Courtesy of Lorraine Clinton Beckstrand – Students who rode with Hy Johnson from the U.S. to the Copperfield School are, (L to R front) Donny Madsen, Billy Collard, Lorraine Clinton and Bobby Johnson. The back three, L to R are Dee Madsen, Grant Collard and the driver, Hy Johnson. Wanda Ruth Jones was absent from this photo. These are the children who "Went Up the Shaft After School". Grant Collard apparently caught the regular school bus from Copperfield down to the Bingham High School in Copperton and back each day.

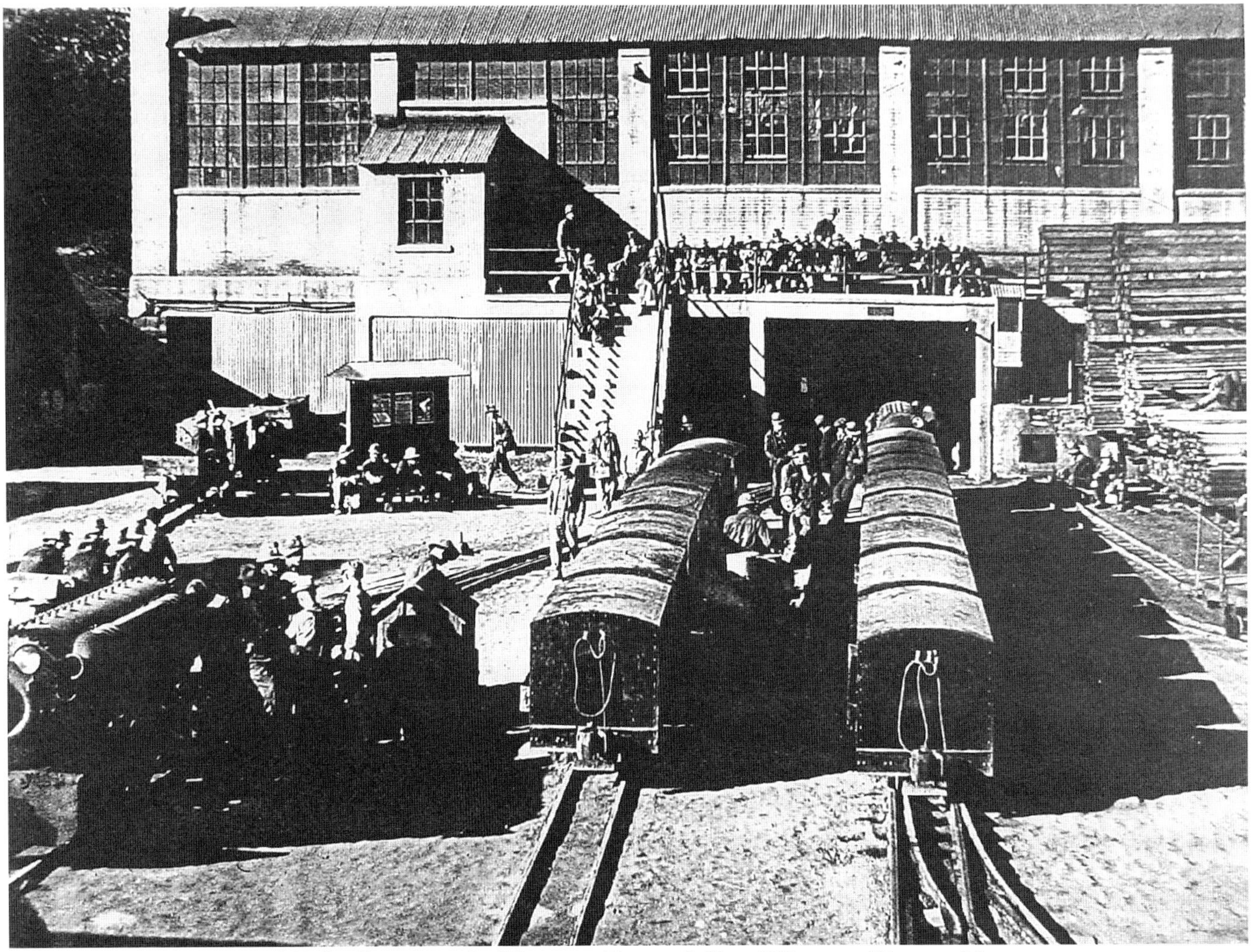

Figure 114 – Portal of U.S. Company's Niagara Tunnel, Dry/Change House Above
From the Gordon Bodily Collection - This photo shows where Lorraine Clinton and other kids from the U.S. started their trip to the shaft station and then up the shaft to get home on some stormy days. They rode in a man-car such as those shown in this photo.

car (they called it a bus). The U.S. was at an elevation of about 7,700 feet, however, and the road was often impassable in the wintertime. On mornings when the snow was too deep and the road not plowed the kids would just stay home.

Sometimes, however, it snowed heavily after the kids had already been taken to school. This presented a problem that the underground miners solved for them. The kids were taken home through the underground mine workings! They walked the quarter mile from the school to the mouth of the Niagara Tunnel that was behind the U.S. Hotel. The miners put the kids onto a man-car that they coupled onto an ore train. The engine then pulled them almost a mile and a half through the dark tunnel to an internal shaft station.

After they arrived at the shaft a miner jerked on the signal rope to have the cage brought up or down the shaft to the station. He then opened the gate and the door and helped the young children into the protective steel cage that was suspended in the shaft by the cable that disappeared into the darkness above them. There was a three-inch crack between the cage and the timbers of the station. As they stepped across the crack they could see into the bottomless void below them. The shaft was over a thousand feet deep. It extended both upward and downward farther than they could see. It seemed to have no top or bottom. And they would be completely dependent upon the hoist cable for their very lives. The kids were a little scared but the miners assured them that everything was alright. Most comforting of all, however, Lorraine knew that her father was at the top of the shaft and was controlling the hoist that would lift them up to the level that led out to their homes.

Figure 115 – A High-Pressure Air Locomotive Just Outside the Portal of the Niagara Tunnel From the Gordon Bodily Collection - This locomotive was used in the U.S. Mines only on the main level to haul men, equipment, timber, etc into the mine and men, ore, and waste out of the mine. This type of locomotive was not widely used in the mining industry. It was very costly to install, requiring a very expensive compressor and extensive piping. Its big advantage, once installed, was that it was inexpensive to operate.

A miner made one more check to make sure the kids were safely enclosed within the cage and then he pulled on the rope to signal the hoist operator to raise the cage. The cage gently started up the shaft but quickly gained speed. They could feel the cage floor pushing up against their feet – it was like they weighed a lot more than they really did. Then the cage settled into a steady upward speed and they were whisked up the shaft. Only a couple of minutes later they felt almost as if they were floating and there was a ticklish sensation in their stomachs as the cage slowed down and then stopped at the shaft station. A miner lifted the gate and opened the door for them. The kids quickly stepped across the open crack between the cage and the strong timbers of the station.

Lorraine's dad, Charles "Blackie" Clinton, was operating the big hoist from the hoist house that was on the surface a hundred feet directly above them. They couldn't see him and he couldn't see them but he knew, from the signals the miner sent him, that the kids were safely out of the cage and ready to walk out of the tunnel. He set the brakes on the hoist, turned everything off, and started walking down the hill to the tunnel entrance to meet his daughter and the other school children.

Figure 116 – High Pressure Air Locomotive on Main Level of U.S. Mine
From the Gordon Bodily Collection - The locomotive in this photo is pulling ore cars to which a man-car would have been coupled when taking the school children to the shaft station.

The miner at the shaft station patted Lorraine on her shoulder and shepherded all the kids to the tunnel right next to the shaft. Then, with his carbide lamp lighting the way, he led them toward the portal where the open air awaited them. They walked through the dark and spooky tunnel for about ten minutes before they came around a curve and saw the light of day ahead of them. Lorraine's dad met them just before they came out of the tunnel at the entrance that they were so familiar with near their homes.

It had been a real adventure for them to have come home from school in such an extraordinary fashion. Very few kids have ever traveled home the same way they did and they were excited to have done it. They would repeat the trip two or three times a year while they went to school in Copperfield and come to accept it without any fear. It was certainly a remarkable way of getting home and an experience they would never forget.

Figure 117 – Miners Doing Repair Work on Shaft on July 4th or December 25th
From the Gordon Bodily Collection - This work was always done on holidays. The school children were hoisted up this or another similar shaft.

Figure 118 – Hoist Main Level to Upper Copperfield (The U.S.?)
From the Gordon Bodily Collection - This is similar to the hoist that Lorraine Clinton's father operated to lift the school children up the shaft from the Niagara Tunnel (Main Level) to the U.S.. Lorraine remembers the hoist room as being much cleaner than this.

THE ACCIDENTAL HANG-GLIDER

With Eugene H. Halverson

Telegraph was the uppermost part of Copperfield, or Upper Bingham in 1938. The old U.S. was a mile or two farther up in Galena Gulch so it wasn't generally referred to as part of Copperfield. Halversons lived in Telegraph just below the sharp turn of the road that went up to Galena Gulch. Their house was the very farthest up the canyon (Bear Gulch) so they were somewhat isolated.

There were no other kids of his age nearby so Gene often entertained himself. When he was eight years old he started building model airplanes. He started out with a balsa wood biplane kit that had a wingspan of eleven or twelve inches. He continued to make small airplanes for two or three years. An advantage of biplanes was that when one crashed there was usually one pair of wings that was still useable for the next plane. With each plane he became more adept at improving their flying ability by such devices as adding adjustable weights to the wings. He made a lot of planes with twelve to fourteen-inch wingspans and he gained a following of several other boys who came up from Copperfield and Dinkeyville to witness the flights.

Halverson's house was on an old mine dump that was a good launching site for his kit-built planes. They could be thrown from thc top of the stairway in his front yard. The house of their nearest neighbor, Marcell Chia, was also built on a mine dump a bit farther down the canyon. The wind currents were usually better at Chia's so Gene sometimes went there to fly his planes.

These planes never did fly much more than a hundred feet or so. Some of them stayed up in the air for a fair while and did some dipsy-doos before crashing but they didn't get much distance. The last kit plane he built was a 14-incher.

After the 14-incher he decided to embark on a grandiose project. He would build an airplane that was bigger than any that they made kits for. He went down to the lumber shop of the U.S. Mine (the Niagara Tunnel) in Copperfield and found some 2-inch wide by ¼-inch thick by 14-foot long pieces of scrap wood that were laying outside in the yard. He picked up as many pieces as he could drag and took them home. He worked in his front yard to build this enormous model. Rather than airplane glue he used bailing wire (also from the lumber yard) to hold it together. The covering for the plane was an ingenious material that most boys didn't often see – his mother's and sister's white silk petticoats and pink bloomers. It took him two or three months to build the huge plane. When finished the airplane had a fourteen-foot wingspan, a four-foot long body and weighed about twenty-five pounds.

There was one critical problem with the plane, however. It wasn't balanced properly. Gene got a large brass bushing from an old, abandoned steam-driven hoist engine at the old Giant Chief mine. The bushing weighed fifteen or twenty pounds – almost as much as the airplane itself. He attached it to the front end of the plane so that its position could be adjusted to fine-tune the overall balance.

A really big airplane, however, required a much bigger, better launching site – one with a vertical lift of a hundred feet or more, an unobstructed takeoff path, and half a mile or more of gliding space. The old Giant Chief mine once again proved to be useful. The dump of the old Giant Chief was only a quarter mile up the canyon from his house. It was a half mile walk along the road and the overgrown trail to the top of the dump but was well worth the effort.

Waste rock from the mine had been hoisted out of the mine and ejected over the hill for twenty years or so beginning in the 1870s or 80s. With each carload of waste that was hoisted out of the mine and tipped over the edge the dump extended farther and farther out from the mine shaft and the waste rolled and slid farther and farther down the hillside. By the time the mine shut down in the early 1900s the waste material had covered all the bushes and small trees for a distance of two hundred yards down the hill and the vertical distance from the top of the dump to the bottom was over two hundred feet. In later years a road had been bulldozed across the top of the dump so there was an elevated point on the very end.

Gene decided that this dump was the perfect spot! The vertical height and unobstructed takeoff path were more than adequate. And for a glide space – he could see down the canyon for miles! If a plane, after takeoff, didn't lose altitude too rapidly and kept to a course down the canyon it might soar for three miles or more!

The big day of the initial flight finally came. His brother Lee and six or eight of the boys from Copperfield carried the plane from the house up to the dump while Gene followed behind with some tools. There was a strong wind coming up the canyon – this would help the airplane take off. He and all the bystanders envisioned this maiden flight in their minds – Gene would hurl the plane off the dump, it would swoop down the hill, gaining speed, and then lift up over the trees and glide away down the canyon like a huge bird. It would fly on and on, becoming smaller and smaller in the distance until it was only a speck in the sky above Copperfield and then they would lose sight of it as it sailed over the open-pit copper mine and went on past the main town of Bingham. Hundreds of people would see it and point at it in the sky above them. There might even be articles in the newspaper of the sighting of the mysterious, silent airplane.

Gene stood on the high point of the mine dump and held the plane up in the air with both hands. Then he shifted it around to determine whether the nose was too heavy or too light. He laid it back down and moved the brass bushing to adjust the balance. He checked the balance and moved the bushing several times until it felt right to him. Then he held the plane above his head with both hands and prepared to launch it.

He wanted to make a short dash forward and throw it off the dump just as a gust of wind came up but it was hard to guess when that would be. It was hard for him to stand steadily upon the tip of the dump while holding the plane that was being buffeted by the wind so the other boys crouched down by his side and helped him keep his balance.

Gene took several quick steps toward the edge of the dump but the wind died down so he stopped short and held onto the plane. He almost fell but the other boys caught him. Then he tried again but with the same results and the boys caught him once more. On the third try, however, there was a sudden, extra-strong gust of wind that caused the plane to lift up higher above his head and he slipped over the dump. The other boys tried to catch him but missed.

Gene held on helplessly as the plane pulled him off the tip of the dump and started flying down toward the trees and the roadway below! At first he tried to run while holding onto the plane but after twenty yards the plane picked up speed and his feet barely touched the ground a couple more times. The surface of the dump was as hard as concrete and very rough – he didn't dare fall onto it. Then the plane lifted still higher and, with him dangling below it, zoomed down the mine dump at what seemed like forty miles an hour! He had no control over the plane and held on for dear life, not knowing what else to do! For a fleeting moment he thought the plane might crash into the trees so he only had to hold on until then. However, by the time the plane reached the bottom of the dump Gene was ten feet above the ground and the plane was gaining altitude and speed with every second.

The other boys, after failing to catch Gene as the plane pulled him over the edge, were stunned to see him still holding onto the airplane and racing at breakneck speed down the dump. Then, as the plane continued to go faster and higher, they realized the seriousness of the situation. They all started to shout "Let go! Let go!"

He finally did let go when the plane was flying over the trees below the dump. He fell eight or ten feet down into the chokecherry trees. The branches broke his fall somewhat but he still hit hard. He was scratched and bleeding and his clothes were all torn. A small tree stump had gone up between one leg and his pants. He had some bruises and limped for several days but he wasn't seriously hurt. The other boys hurried down to help him and were relieved to find him in as good a shape as he was. It was a good thing he let go when he did. If he had hung on until he was over the road he would have fallen a hundred feet onto a hard surface and could have been killed.

After Gene let go the plane went straight out over the trees, crossed the road and flew over his house. Then it climbed high in the air and seemed, for a long time, to float lazily on the wind that was coming up the canyon. After several moments it slowly dropped downward and floated backward as it got lower. It then gently began going down the canyon again. It appeared to be flying like a feather. But then what looked like and flew like a feather crashed into Chia's garage with a loud bang! It hit so hard that the body and wings of the airplane were broken so badly that they were only good for kindling wood. The brass bushing came loose from the rest of the plane and made a big hole in the top of the garage. The length of the flight, from the mine dump to Chia's garage, was about half a mile.

There were no grownup witnesses to the mishap but the boys were afraid of being held responsible for the damage to Chia's garage. And what might the heavy brass bushing have done to Chia's car? They gathered what was left of the plane and hid out until they found out that the car was away somewhere and hadn't been in the garage. Gene saw Mister Chia look at the hole in his garage many times and Gene always wondered if he ever knew who did it. It would have taken a huge monster to lift the bushing and throw it that high and with such force! Neither Gene nor Lee ever told their Mom or Dad about Gene's hang-gliding experience.

This hang-gliding flight, though accidental, was certainly the first in Bingham Canyon and probably the first in the state of Utah. It wasn't a particularly long ride (just a bit longer than the length of the Giant Chief mine dump) and it wasn't reported in any newspapers. But it has lived on in the memories of all who witnessed it and of many who heard the story. It is remembered by Gene as one of his outstanding lifetime experiences.

Figure 119 – Upper Telegraph w Giant Chief Dump, Mayberry Mine, Emil Bodmer's House
Photo Courtesy of Max Ivie and of Eugene H. Halverson – The Giant Chief mine dump is the long dump that extends from the right edge two-thirds of the way up the photo. Gene's hang-glider flight (The Accidental Hang-Glider) began at the top of the extreme left edge of the dump and he fell into the trees at the bottom. The Halverson and Chia homes are behind the large dump of the Mayberry Mine which is halfway up and on the right side of the photo. A building of the Mayberry Mine is visible just above the right end of the Mayberry dump. Karl John's mine (The Secret Mine of Telegraph) was in the area that is covered by the dump and the mine building. The ore chute to which Karl wheelbarrowed his ore is in the shadows near the left center of the photo. Karl pushed his wheelbarrow from the right side of the photo over to the main road, up the road to near the water tanks and then to the left back to the ore chute – in the dark of night.

Emil Bodmer's house, where he lived with his mother, is near the center of the photo between the water tanks and the Mayberry Mine. Emil was a bachelor who worked for the Utah Copper and was also a Dollar-a-Year part-time game warden. Emil figured in the story "One Hunter Parties While Other Suffers".

THE SECRET MINE OF TELEGRAPH

With Eugene H. Halverson

Karl John (Johan) shoveled the last of the broken ore into his wheelbarrow. The ore was heavy and a full shovelful of the dark ore was almost more than he could lift. He was digging a trench along the 5-foot wide vein and was now about 7 feet below the surface. The ore was broken and weathered so that it didn't look very spectacular but it was worth a lot of money. It was composed mostly of galena, the lead mineral that often carries silver – and this ore was very rich in silver. This was what prospectors dreamed about! It was very high-grade ore that he could make a fortune from!

Most miners would delight in such a find and would ship as much of it as they could as quickly as they could. They would be paid handsomely for this spectacular shipment and in only a couple of weeks they would have a check in their hands with which they could buy a new car and have money to spare! They could show this check to their wives and family and friends. It would be something to brag about. But not Karl – he was too wise. He kept it a secret from everybody – or at least he tried to.

Karl held a lease from the U.S. Company and he shipped the ore to their concentrator down in Midvale. The Company then sampled Karl's ore and weighed it before running it through their concentrator and smelter. He would be paid according to a set schedule for the values of lead, silver, and other metals in the ore and how much it weighed. The Company kept very good records of all the shipments from the various leases. If a particular lease was of too high a grade then the Company would rather mine it themselves than to have the leaser grow rich off their ore. The Company could cancel the lease or not renew it at any time and Karl knew this very well. So Karl devised a system for fooling the Company.

There were three major requirements for his strategy. First, the Company's assays from his ore must not be too high. Second, nobody else should ever see the vein or the high-grade ore that he was mining. And third, nobody else should ever hear about it. The problem of the assays could be solved by mixing the high-grade ore with low-grade rock before he put it into the ore chute. He could hide the vein by making sure nobody saw him working on it or saw him taking the ore to "his" chute in his wheelbarrow (Karl didn't own the chute but he was the only one who used it). As for talking about it, he was the only one who knew about the vein and he would absolutely never talk about it to anybody. And so he was safe (or so he thought).

Karl adopted a secretive life style in order to mine the ore, dilute it, and get it to the ore chute without being seen. He went to the mine that was only a few yards from his house and came home on an odd schedule that was least likely to be observed by anyone else. Diluting the ore was no problem – he could just mix in loose rock from next to the vein or waste rock from the dump next to the ore chute -- when nobody was watching of course. Getting the ore to the chute was, however, a real challenge. He had to push his wheelbarrow from the trench where he was digging it, up and across the road, up the short grade to the top of the ore chute, and then tip it in – without being seen. Luckily enough, Karl's trench was less than a hundred yards from the ore chute. Karl must have solved this problem by making his wheelbarrow trips at night when everyone else was sleeping.

Figure 120 – Karl John's Secret Mine
Photo Courtesy of Eugene H. Halverson - Karl's mine (trench) was just under the dump of the newly constructed Mayberry Mine in the foreground. Karl's house was about where the tracks are. The ore chute was across the road just below Emil Bodmer's house which is on the right side of the photo. After the U.S. took over the lease they built and started up the Mayberry Mine which headed in the same direction as Karl's trench. It was apparently exploiting the same vein of high-grade ore that Karl had discovered and mined for years.

The downfall to Karl's secret started with an automobile/sleigh accident down in Frog Town. Fred Hoyne, the son of the U.S. Company's superintendent, was driving up the main road quite fast just as two young boys, Gene and Lee Halverson, came from a side street on their sleigh. There wasn't time for the "lookouts" to warn Gene in time for him to stop. The sleigh slammed into the front side of Fred's car, injuring both of the little boys. Gene's hip was broken and Lee was knocked unconscious and bloodied but he wasn't seriously injured. Gene and Lee's Dad, Harvey, was out of work at this time because he had miner's consumption (silicosis) and so could no longer work underground for the U.S. as he had previously. This accident led to Harvey being rehired by the U.S. for the job of maintaining the air compressor plant that was outside the mine. The former compressor operator was leaving and so Harvey was also provided with his house.

The Halverson family moved to their "new" house that was on top of a mine dump at the upper reach of Telegraph. Gene was in the hospital in traction for several weeks but finally was able to come home. But he could hardly walk, even with crutches, and had lots of time to just sit outside the house, day after day, through a summer and a winter. He missed a whole year of school because of this injury.

While sitting in his front yard, on top the mine dump, Gene was kind of like a big bird with a bird's-eye view of all the proceedings in his proximity. He soon began to notice the strange antics of Karl John and started to watch him. He could tell that Karl wasn't fond of being watched even by an eight year old kid and that it bugged him but Gene continued to sit and to watch. Karl began watching Gene as much as Gene was watching Karl. After a few months Gene figured out where Karl was going on his clandestine forays so one day he followed Karl and sneaked up on the mine while Karl was working. But he wasn't stealthy enough and Karl caught him. Gene was afraid of what Karl might do to him and Karl was afraid that Gene might tell somebody about his mine.

The two of them sat down and had a serious talk. Karl immediately began to question Gene. "What do you know and who have you told? What about your parents – do they know?" Gene said he had known that Karl was hiding something and now he knew what and assured him that nobody else knew anything about the mine. Karl asked him to keep quiet or he (Karl) would be sent away. So Gene promised. Lee also found out about the mine but he also kept quiet about it. Gene patted himself on the back for finding out Karl's secret.

Neither Gene nor Lee were ever able to catch Karl pushing his wheelbarrow from his trench to the loading chute. They did discover, however, that if the ore chute was full, with no more room for his ore, then Karl would push his wheelbarrow as much as a hundred yards into an inactive mine that was next to the chute, dump it there, and leave the ore hidden. Sometimes he had quite a few wheelbarrow loads of ore hidden in that old mine before the chute was emptied again. After his ore from the full chute was trucked to the railcars down by Fox's Switch then Karl would go back into the mine, shovel all his hidden ore back into his wheelbarrow and take it out to the now-empty chute.

Karl's secret mine was well hidden. The vein didn't crop out on the exposed surface -- it was 5 or 6 feet under the stream gravels and couldn't be seen by a casual observer unless he almost stepped into the trench. Gene thought, later, that Karl had dug down to bedrock looking for placer gold and hit the vein accidentally. Four other shafts were dug within a few feet of his lease but all missed his orebody.

Karl worked the mine for years with a pick, shovel, and wheelbarrow without being found out by the U.S. Company. He obviously made good money from his ore shipments but never overdid it to the point of losing his lease. He is thought to have retired a rich man but nobody knew for sure. He was in his 50's or possibly early 60's when he retired. Before leaving he covered or filled in his ore trench with waste rock and dirt.

After Karl had retired and moved away, the Heinicke brothers began scraping the creek bed down to bedrock and they found Karl's orebody. But greed overtook them. They shipped the high-grade ore without diluting it and the U.S. Company shut them down after the first shipment.

Few if any people knew Karl John well. His last name might not be John at all – that might be his middle name. Everyone thought he was wealthy and he might have been at one time because he had boxes of old German coins and paper money. If a kid asked him for some money he would give him some. There were always kids leaving his place with pockets full of money. Another memorable thing about Karl was that he had his own forge and anvil. The kids loved to watch him as he heated picks and bars until they were red hot and then hammered the ends to sharp points on his anvil.

Aside from these activities, however, Karl kept to himself and was a secretive old cuss. He was never mean or ornery but he had few if any visitors or friends and few people ever stopped to talk to him. He never seemed to smile except to laugh at himself. Another thing about Karl was that he seldom went anywhere; Gene never saw him in any of the stores in Copperfield or Bingham. He didn't own an automobile. The only thing he owned that had wheels on it was his wheelbarrow. Once in a while a taxi or car would pick him up and he would leave but apparently he went straight to Salt Lake City -- that was where he said he was going when he retired. He probably had friends or family there.

UNDERGROUND WITH NO LIGHT

With Virgil Scorzato and Max Sluga

Christmas was approaching so Virgil and Max set off up the mountain from Highland Boy toward Sunshine Peak to get a Christmas tree for each of them. They were sixteen years old and full of energy. They climbed up the steep hillside through the snow until they reached the ridge between Highland Boy and Eppus Canyon (the left-hand fork of Middle Canyon). Then they worked their way around to the ridge overlooking Middle Canyon itself. They hadn't intended to go this far but they hadn't seen any good-looking Christmas trees yet. They dropped down the mountain farther and farther until they were most of the way to the bottom of Middle Canyon where they found some large pine trees. Each of them found a tree with a good-looking top. They each climbed a tree and chopped off the top. Max's first tree broke when it hit the ground so he had to cut another one. Now they each had a Christmas tree about ten feet tall but were faced with the problem of how to get them home.

It would be a long, difficult hike to go back the way they had come and would be impossible if they carried and dragged the trees. But they were now only a short distance above the caretaker's house by the Utah Metals Tunnel. Virgil lived in Highland Boy and had been through the tunnel several times. He figured that going through the tunnel was the much easier way to get back home. So they dragged their trees straight down the hill to the bottom of Middle Canyon and to the nearby entrance to the tunnel. The tunnel was 2 ½ miles long -- it went all the way through the mountain and came out in Highland Boy.

There was a pushcart by the entrance onto which they loaded their Christmas trees. There were no lights in the tunnel and they hadn't brought a light of any sort. They looked around and, in a nearby shack, found a gallon jug of alcohol. They broke the top off the jug and lit the fumes with a match. It burned almost like a lantern. With the "lantern" set on the pushcart along with the trees, they pushed the cart into the tunnel and proceeded toward Highland Boy. The cart rolled smoothly along the track and it was a very gentle uphill slope. This was much easier than hiking back up and over the mountain to Highland Boy and besides that they had their trees which were almost no burden now.

All went well until, halfway through the tunnel, with more than a mile still to go, the glass jug broke and all the alcohol ran out. There was a short period of bright light when all the alcohol was burning but it soon went out. Then it was pitch black! In all directions. The Middle Canyon entrance was more than a mile behind them.

Luckily enough, Virgil had been through the tunnel before and knew that if they kept to the tracks they would get to Highland Boy. So they both held to the back end of the pushcart and began pushing on through the inky blackness. They each walked on one of the rails – one foot right in front of the other like trapeze artists walking a tightrope. They safely passed two or three crosscuts and the water shaft without being able to see them. The pushcart moved easily now so they knew that the tunnel was sloping downward ever so slightly toward the Highland Boy entrance. They looked forward every few minutes hoping to see the pinpoint of light at the entrance but over and over again they saw absolutely nothing.

After almost an hour of holding to the pushcart and walking on the rails as they pushed the cart, they finally looked ahead and saw the welcome light of day. It was late afternoon when they pushed the cart out from the tunnel and breathed a happy sigh of relief. Even with having lost their "lantern" they had made it back to Highland Boy far sooner and easier than if they had hiked back over the mountain. And they had their trees.

Tunnel Entrance—Utah Metal Property.

Figure 121 – Middle Canyon Portal of Utah Metals Tunnel in About 1909
From "Bingham, The Greatest Copper Camp In The World" 1909 – The tunnel was 2 ½ miles long and went all the way through the mountain from Middle Canyon to Highland Boy. A 2 ½ mile walk through the tunnel, on solid ground that was nearly level, with the Christmas trees riding on the pushcart, was far easier than carrying the trees up over the mountain -- especially in deep snow.

But the adventure wasn't totally over yet. Virgil had lost his wallet somewhere in the tunnel. It had his driver's license and a few bucks in it and he wanted to get it back. The next weekend Virgil and his friend Bob Christie used the "pump-car" to go all the way through the tunnel and back while shining a flashlight along the track in front of them but didn't find the wallet. The waterman went through the tunnel every day on a bicycle but he never saw the wallet. It was a month later that Virgil, by himself, walked back into the tunnel using a flashlight. Lo and behold, there was the wallet, lying open on a tie, untouched, with water and mud on either side! The chances of it falling right on top of a tie the way it did were mighty slim.

Virgil was extremely happy to get his wallet back. The whole adventure had ended well. They had found their Christmas trees, made their way through the pitch-black tunnel with the trees and Virgil had found his wallet. Years later both Virgil and Max looked back with fondness upon this exploit.

CHAPTER SIX

UTAH COPPER STORIES

There must be dozens or hundreds of interesting stories about happenings at the Utah Copper. Following are a few that I have heard. It is lamentable that many others have been lost by not being written down before the passing of the old-timers who experienced or knew about them.

FALSE TEETH IN THE WATER BUCKET

The Utah Copper open-pit mining operation has, since its earliest days, required the stripping of waste rock from the periphery of the orebody. The waste is hauled away from the mine and dumped into the side gulches of Bingham Canyon or around the mountain into gulches that drain directly into the Salt Lake Valley. For the first eighty years or so the waste was hauled in side-dump rail cars; the use of large trucks to haul the waste is relatively new. In the earliest years of the operation the distance from the mine to where the waste was dumped was less than half a mile. However, as the mine was enlarged the gulches nearest to the mine were filled in and the waste was hauled farther and farther – it was a major operation.

Hundreds of men were used to maintain the railroad on the waste dumps and during the summer especially they needed drinking water. The track gangs, of about fifteen to thirty men had their own water buckets. The locomotives that hauled the waste cars out to the dumps often had their own water bucket also. However, there were dozens of other men such as trackwalkers and timekeepers who didn't carry their own water and when they got thirsty they would sometimes get a drink from a track gang's or locomotive's bucket.

One hot summer day a timekeeper was especially thirsty so he climbed onto a locomotive to get a drink. The locomotive engineer was an older guy with false teeth who was pretty independent and hard to get along with. The timekeeper climbed onto the engine and went straight to the water bucket. The top of the bucket had a wooden cover that was divided in two; the two halves were connected by a rubber hinge so that one side could be tipped up to either fill the bucket with water or to withdraw a dipperful to drink. There was a notch in this side that the handle of a long-handled dipper fit through and the dipper itself rested in the bottom of the clear, cool, sweet water that the timekeeper was after. The water itself couldn't be seen because of the wooden cover. The timekeeper grabbed the dipper handle with his right hand, tipped the wooden half of the cover up with his left and lifted the dipper of water to his mouth. Just as he was about to start drinking he looked and saw the engineer's false teeth sitting in the dipper! It almost made him gag! The engineer had put them there to soak. The timekeeper stuck the dipper back into the bucket and started cursing the engineer. The engineer said that was his water bucket and he would do whatever he wanted to with it. After a few more bitter words the timekeeper stormed off the locomotive, still thirsty. Word got around about the incident and after that nobody would trust the water from that locomotive. But the engineer didn't much care -- he just kept on doing his job of running the engine with maybe a bit more of a sour look on his face.

LIONS IN THE NIGHT

The higher-elevation levels of the Utah Copper open-pit mine had to be excavated farther back away from the center of the mine as the bottom of the mine was deepened and expanded. The waste material from the upper levels was hauled away from the mine in side-dump waste cars. There was limited space for waste dumps on the upper East Side of the mine so the dumps were continually extended farther and farther out from the mine proper. Much of the country adjacent to the mine was prime deer habitat. The upper East Side was especially good deer country to which many deer migrated in the winter. It included The Big Rocks and Fortuna as we called the area. It was possible to see over a hundred deer in one day of hiking or hunting there in the winter. Since lions prey primarily on deer it was only natural for there to be lions in the same area. The waste dumps extended into this prime deer area for perhaps thirty-five years until the waste dumps finally covered it completely. The train crews hauling waste out to the area saw deer almost every day; coyotes and an occasional lion were also spotted.

The trains of waste cars would be pushed out to the dump area and the exact spot for them to tip the cars over would be assigned to them by a "dumpman". The cars couldn't be emptied just anywhere because they might "plug" the dump and make it unusable until the dozer and track gang could prepare it for more dumping.

The dumpman generally stayed out on the dump throughout his shift and the mine operated three 8-hour shifts per day. There usually was an iron shack for the track gang to eat their lunch in and for the dumpman to use for shelter but sometimes the shack couldn't be used because it was too far from where the cars were being dumped. During the afternoon and night shifts the dumpman would be the only person out on the dump except for the short intervals when trains would come out (once an hour or so), dump their cars in just a few minutes, and then return to the shovel to be loaded again.

It could be a lonely job for a dumpman to be by himself in the dark out in the hills where wild animals were roaming. One dump, up toward Q-Level, became extremely frightening for the dumpman one night when he heard some lions snarling and yowling. He didn't know whether they were fighting or just what but the sounds were bloodcurdling and he was very afraid that they might come after him. As soon as a train came out to dump he climbed onto the motor and told the engineer and brakeman about the lions. Rather than stay out on the dump by himself the dumpman started riding the motor out to the dump and back whenever possible. The screeching and yowling by the lions was repeated intermittently for several nights. It wasn't always feasible for the dumpman to ride the motor, however, so he had to remain out on the dump part of the time and was extremely nervous when he did so.

One night the train crew came up with a bright idea. They cut two round holes about six inches apart in the bottom of a cardboard box. Then, as quietly as possible, they ran the train out onto the dump and stopped short of where they knew the dumpman was laying on the ground. The brakeman then put a lantern with a red light inside the box and sneaked up toward the dumpman until he was twenty yards away. Then the brakeman turned on the red light so that it shined through the two "eye" holes and he started growling like a lion. The poor dumpman was laying face down on the ground. When he heard the growling he jumped to his feet, saw the two red eyes of the approaching lion, and started running as fast as he could over the rough ground toward where he hoped the engine was. Luckily enough, he made it safely to the engine and quickly climbed aboard. The engineer and the brakeman almost went into hysterics as they laughed about the trick they had played on the dumpman.

Nobody ever saw the cats in the daytime so they couldn't be positive whether they were lions or only bobcats. But they sure made a lot of noise. Most likely they were mating rather than fighting. In either event they almost certainly never would have attacked a man. But all the dumpmen who worked on that particular dump for weeks afterward were very scared or even terrified of having to work out there by themselves in the dark.

A BULLDOZER LOAD OF DEER

The hills above the West Side dumps of the mine had as many or even more deer than the East Side. In addition to deer there were quite a few elk on the West Side. Dry Fork, along with the rest of Bingham Canyon, had been closed to hunting for years and the deer just kept multiplying. In addition, when the hunting season opened in Middle Canyon and other areas to the west of Bingham Canyon, the deer would make a beeline for the safe haven of the closed area. On opening day when hunters made their way up to the Middle Canyon/Bingham Canyon divide they were usually greeted by fresh deer tracks that led up, over the ridge, and then down into the Bingham drainage. The result was that Dry Fork (along with Markham and Freeman gulches) probably had the greatest concentration of deer in the whole Oquirrh mountain range and probably in the whole state of Utah.

The waste dumps extended across the lower parts of Markham and Freeman Gulches and were reaching toward Dry Fork in 1953, the first year that deer hunting was allowed in the area. The track gangs and other men who worked on the Markham and Freeman dumps began carrying their rifles to work with them. It was a simple matter to walk a short distance up the mountain before or after shift or even at noontime and shoot a deer. The deer were so plentiful that some of the men who didn't have licenses borrowed rifles from the legal hunters and bagged a buck for themselves.

The more difficult part of the hunt was to drag the deer down to the road where cars could be driven. A dozer trail ran from H-dump down to G-dump. The workers (and hunters) could drive their cars to G-dump. The hunters had managed to get their deer to the upper part of this dozer trail when along came a bulldozer going down the hill. The hunters waved to the operator and he stopped for them. Since he was going down the hill anyway he figured he might as well help the guys out by hauling their deer on the dozer. The dead deer were loaded onto the dozer wherever they could fit. Deer were laid on top of the cab, across the hood, draped over the blade, and then stacked on top of one another. It was amazing -- twenty-seven deer were put onto that bulldozer!

It was then a simple matter of the dozer carrying all the deer down to the road on G-dump. The hunters took a shortcut down the hill and through the short railroad tunnel while the dozer had to take a longer route. But then the guys saw the Fish and Game wardens on the other side of the tunnel checking all the cars of hunters and their deer. Rather than wait for the dozer and going through the checking station with untagged deer and no licenses, the track gang hunters all left for home.

The reaction of the game wardens when they first saw the bulldozer piled high with deer hasn't been recorded but it must have astounded them. Nobody could have ever before seen or even imagined a bulldozer carrying twenty-seven deer! The dozer operator was caught redhanded with over twenty untagged deer and without a license himself!

The game wardens took all the deer to a meat locker. The operator was held responsible for all the deer that he had been illegally transporting and for paying a $20 fine for each deer. They told him that he would be relieved of responsibility for each deer he found an owner for. The hunters got together and agreed to accept responsibility for the animals. They were allowed to go to the locker and pay $20 each to retrieve their deer. Somebody made off with one deer without paying for it so the operator ended up having to pay $20 without even getting a deer for himself. It was also said that the bulldozer was confiscated but it never left Kennecott property. The gang boss said he never told the dozer operator that he could use the machine to haul the deer so the dozer operator was fired from his job.

The affair involved the Utah Fish & Game and more than two dozen hunters who were employees of Kennecott Copper, and Kennecott's bulldozer which had been used to transport illegal game. The final result wasn't as bad as it could have been. The hunters relieved the dozer operator of the fine, the operator was reinstated in his job and the dozer remained the property of Kennecott. The Fish & Game put up signs declaring Markham and Freeman Gulches closed to hunting but Dry Fork remained open.

The biggest question was whether the whole deal was a lesson about how the law should be obeyed or whether it was a big comedy. Most people couldn't help laughing at the image of the dozer operator running his machine with twenty-seven deer stacked on it while the game wardens stared at the sight with their mouths open.

ODD JOB ON A STEAM SHOVEL

My Dad told me of a comical incident that happened at the Utah Copper mine many years ago. A new man was to start working on one of the old Standard steam shovels so the level foreman took him to his new job. They walked out to the shovel where the rest of the crew was already at work. The foreman was in a hurry so he pointed up to the shovel runner and told the new man that the shovel runner would show him what to do. The shovel runner waved back to him, acknowledging that he would take care of the new man. Then the foreman left to take care of another job.

The next day the foreman went to check how the new man was doing. As he approached he was greeted by the most ridiculous and astounding sight he had ever seen! The new man was standing on top the shovel with a big sledge hammer and he was black with soot. As the foreman watched, the man swung the sledge hammer and hit the smoke stack with a loud "Clang!" upon which the shovel runner blew air up through the smoke stack and soot was blown out all over the man wielding the hammer. The procedure was repeated several times with the same result of soot being blown on the unfortunate workman. The foreman hastened up to the shovel and demanded to know what was going on. The new man explained that the shovel crew had told him that part of his job was to keep the soot from clogging up the smoke stack and he was to do this by hammering on it as the shovel runner blew air through it. Needless to say, the smoke stack didn't need cleaning but the shovel crew had a big laugh at the expense of their new crew member.

AN UNEVEN BATTLE WITH A SKUNK
With Bob "Trapper" Kemp

All kinds of wild animals often made their way onto the waste dumps which extended out into the hills next to the open-pit mine. Track gangs of about fifteen men worked on these dumps repairing, moving, and extending the railroad tracks; they often saw the wild animals. A trackshifter was a necessary piece of equipment for a track gang – it was used to move the track out to the edge of the dump after the trains had dumped as much waste rock as there was room for (it had become "plugged"). A preferred item for every track gang was a shack where they could eat lunch and get out of the weather if necessary. This combination of a wild animal, a track gang, and a shack led one day to a great adventure or happening.

Bob Kemp was only sixteen or seventeen years old one summer when he was working on the #3 Gang for Don Heilmann. The gang was on C-dumpline on the East Side of the mine. They kept their lunches in the nearby carwhacker's shack and ate there at noon each day. George (Yobo) Borich was the shifter operator and he had a habit of running the shifter out to the job and then letting one of the men on the gang practice operating the shifter while he went to the shack for a rest.

On this never-to-be-forgotten day Bob was given the job of pulling bad ties from under the track while the rest of the gang was working quite some distance away. While working he saw a guy come running out of the metal shack and then run back in. This happened two or three times. Finally he walked over to the shack to see what was happening.

At the door to the shack Bob was greeted by the horrific smell of a skunk (actually a little spotted civet cat). The air inside the shack was saturated with the skunk's pungent odor. The man who had been running in and out of the shack was Yobo the shifter operator. And now he had the skunk cornered in a small space under the shack next to a crack in the metal and was trying to kill it with the stove poker. Apparently Yobo had been quietly resting in the shack when the skunk came in through the crack next to the floor. During the encounter the skunk had sprayed Yobo who now figured that he had nothing more to lose and was determined to dispatch the skunk at all costs.

Bob started carrying the lunches outside where they wouldn't be ruined by the stench. But Yobo told him to forget the lunches – for him to stay inside and watch what the skunk was doing while he ran outside to the back of the shack where he could reach the skunk better and beat it to death with the poker. If the skunk turned to spray Yobo again then Bob was to holler and warn him.

The iron shack was resting on some 2x4s or pallets so that it was three or four inches off the ground. The iron of one corner of the metal shack had been bent and a weld ripped so there was a 2-inch crack next to the floor. The skunk was "trapped" under the floor and Yobo (and now Bob) could barely see it through the crack. By laying down on the ground outside the shack, shutting one eye and squinting with the other one up to the opening under the shack he had a pretty good view of the skunk and with some effort he could jab it with the poker and hopefully kill it.

While Yobo was poking at the skunk Bob saw it turn its rear end toward Yobo and he yelled a warning. But Yobo could also see the skunk and he said "No! I can see his eyes!" Then the skunk sprayed him again! There was no doubt but that the skunk had turned its weapon toward its tormentor. Yobo wore his hair long down over his forehead. The spray caught him on the forehead and up into his hair. This chemical attack was so strong that he started gagging and couldn't breathe. He rolled around in the dirt and wiped his hands across his head in a desperate effort to remove the malodorous liquid. Bob ran outside the shack and saw Yobo on the ground desperately gasping for air. He grabbed the water bucket and poured the water over Yobo's head and face. After that Yobo regained his breath and came out of it OK.

At this point Yobo (and Bob) gave up. Prior to this moment they had fought a good battle against the skunk but had been outclassed from start to finish. The skunk was uninjured and the two men reeked of its "perfume". But now, with its full-force onslaught on Yobo's head from less than two feet away, the skunk had definitely won and the war was over. Bob wasn't sure what finally happened with the skunk but he and Yobo quit the field and went off to lick their wounds. Bob had never been sprayed directly but the air inside the shack smelled so bad that his clothes and especially his hair absorbed the smell which clung to him as if he had been sprayed. Yobo smelled so horrible that no words can describe it.

At lunchtime the rest of the track gang came to the shack to eat. They were dismayed to find their comfortable refuge from the hot sun virtually unusable because of the terrible smell. They picked up their lunches and went to another nearby shack. Then they discovered that all except two of their lunches were unfit to eat – these were the two lunches that Bob had carried outside when he first reached the shack. When the other men opened their lunch sack or bucket and examined its contents they found that the smell had penetrated even the buckets and permeated all the food so that they couldn't force it down. They disconsolately spent the rest of the lunch break trying to rest on empty stomachs. There was considerable questioning and cursing but also some joking about Yobo's wisdom in starting to fight with the skunk in the first place. That very day Bob Kemp was given the nickname of "Trapper". He was known to friends and acquaintances by this name for the rest of his life.

But Trapper's adventures for the day weren't yet over. He lived in Midvale and had ridden to work with eight or ten men in Bunnel's van. Bunnel didn't want Trapper to stink up his van and the other occupants also refused to let him ride inside with them. A partial solution was for him to take off his levis and shirt and wear some coveralls that one of the others loaned him. But this wasn't good enough. He still smelled too bad. His whole body, especially his hair, reeked of the terrible smell. And he was still wearing his boots, which he couldn't do without until after he got home. The final solution was for him to ride the twenty miles home on the outside of the car. The right-front fender offered the most secure position so that is where he rode, sitting atop the fender, holding onto the headlight and with his feet braced against the bumper. Luckily enough it was summertime so the ride home wasn't as bad as it might have been. The wind blowing through his hair at 30 miles an hour was almost relaxing. No doubt some people in Midvale caught a whiff of the telltale smell as he rode by but he was gone before they could say anything.

When he arrived at his home the guys stopped to let him off and, with relief and a chuckle, waved goodbye. Trapper walked around behind the garage to where there was an irrigation ditch and a fire barrel. When he didn't come directly into the house his mother realized something was wrong and came out to help. She brought a big wash tub and put a lot of homemade tomato juice, a couple gallons of milk and some laundry soap into it. Trapper took off all his clothes and threw them into the fire barrel. His boots were unsalvageable and he threw them into the nearby canal. He first washed himself off in the irrigation ditch and then he bathed in the contents of the tub. His hair was so bad that he had to wash it several times over. Finally, he rinsed off with the garden hose. By this time he was reasonably odor-free.

It took only a few days for the story to make the rounds of the mine and thus the name and legend of "Trapper" was created. Yobo Borich, the central figure in the whole episode, escaped without lasting damage to his image. Before quitting time he climbed onto his trackshifter and left for the shifter shed. The welcome he received there and how he got home aren't known. Maybe he just parked the shifter, walked to his car and drove home with nobody being the wiser. The track gang was lucky enough to have the alternate shack to use. The carwhacker's shack was unusable until the next Spring and even then only after they threw gasoline into it and set it on fire so that the wooden benches were burned up.

CHAPTER SEVEN

THE MINE CONTINUES TO GROW, COPPERFIELD MINED AWAY

After the vehicular tunnel was built and the old road between Bingham and Copperfield was severed then the threat to the town of Copperfield moderated. But this reprieve was only temporary. The mine continued to grow and Copperfield was still blocking its way. Development of the south side of the pit was forestalled for several years but then, in one large gulp in 1958, the town was demolished and the mine proceeded to swallow the very ground upon which the town once stood.

CONTINUED GROWTH OF THE MINE

The shape and location of the Utah Copper orebody relative to Copperfield, Highland Boy and the City of Bingham were such that the three communities could, if absolutely necessary, be left in relative peace for some time even though the mine continued to grow laterally to the east and west and in depth. The slopes were steepened in some areas to allow for this reprieve.

Highland Boy was the most critical of the three areas in regard to the need for mine expansion. But the Anaconda Company owned a significant portion of Highland Boy. The City of Bingham (including private property owners) owned most of the property in the main canyon north of the mine. The U.S. Smelting, Refining & Mining Company and some private parties owned property to the south (including some of Copperfield).

An unanticipated factor in the status of Highland Boy and the City of Bingham came into play in the mid and late 1960s. The underground North Ore Shoot was discovered by Kennecott to the north of the pit. The Anaconda Company then began drilling the same orebody on their side of the property line. This discovery probably hastened Kennecott's purchase of property in the City of Bingham and of Anaconda's Carr Fork properties.

After the purchase of all the properties around the mine then Kennecott was free to "catch up" with expansion of the mine in areas where waste removal and other activities had previously been deferred. One of the first requirements was to demolish all the buildings in Bingham Canyon. After that the mine could be expanded in the most efficient manner.

LAST OF THE TOWN MINED OUT

The roadway between Bingham and Copperfield was severed shortly after the vehicular tunnel was completed in 1939. The E-Line Bridge and the nearby buildings up to but not including the Pan-Hellenic grocery were demolished. The mine then expanded its levels outward to the east and south so that the Observation Point on lower Main Street was right on the edge of the pit. After this brief expansion in

Figure 122 – Demolition of The Circle, Schoolhouse and Main Street in Nov. 1958
Photo Courtesy of Shirl and Isabell Scroggin – The skeletal walls of a two-story duplex in the back part of The Circle are in the right side of the photo. The mostly-demolished remains of the Copperfield Grade School and the bathrooms/heating plant are in the center of the photo. The foundations of some houses that used to be on Main Street are on the left. Concrete retaining walls are still present where the Terrace Heights were located. Roadways and power poles are almost the only things left of Dinkeyville.

1939-1941 of the mine into the lower part of Copperfield; then mining in the "Copperfield Corner" came to a near standstill. The parts of the town above the Observation Point and the nearby Pan-Hellenic Grocery were safe for about seventeen years.

The pressure for mining out the town of Copperfield continued to grow however. The U.S. Company owned considerable property and mineral rights in the area and was actively mining underground through the Niagara Tunnel. So Kennecott bought out the properties of the U.S. Company in the Bingham and Lark areas. This buyout along with that of the remaining few private property owners opened the way for mining out the entire town. The end came quite quickly to

Figure 123 – Demolition of the Schoolhouse and Main Street Houses in November 1958
Photo Courtesy of Shirl and Isabell Scroggin – The Schoolhouse is in the center of the photo. The main part of the school that was built in 1918-1919 is on the left and the flat-roofed 1926 addition is on the right. The "alley" between the parking lot (on the right edge of the photo) and the flat-roofed addition is clearly shown. In the lower right foreground are the remains of the end-most one-story duplex of the Lower Terrace Heights and the adjacent concrete stairway. Concrete foundations and partial walls of the houses along Main Street, including the U.S. Hotel are shown. The dump of the U.S. Mine (Niagara Tunnel) is still present.

Figure 124 – Demolition of Upper Main Street and Jap Camp, November 1958
Photo Courtesy of Shirl and Isabell Scroggin – The houses of Tony Jaterka (note the apple tree still standing on the bottom left of the photo) and Rex Graham have been demolished; Nick Golish's is still standing. Fox's Switch looks much the same as in the 1940s. The lowermost houses of Jap Camp are in the early stages of demolition; those on the extreme right have been torn down. The uppermost houses appear to be untouched.

Figure 125 – Demolition of the Lower Terrace Heights Two-Story Duplexes In 1958
Photo Courtesy of Shirl and Isabell Scroggin – The first duplex, that of Walter Leatherwood and Burl Barton in the 1950s has already been torn down so that only the concrete foundations remain. All the houses of Terrace Heights were torn down shortly thereafter. These two-story duplexes were built in 1927 so they had a lifespan of thirty-one years. The foundation and supports of the U.S. Hotel are on the left of the photo, next to the U.S. Mine (Niagara Tunnel) dump. This photo also shows part of the roadbed of the Shay Track of the Copper Belt Railroad (on the lower left corner of the photo and extending past the fire hydrant shed) which dates from about 1901.

Copperfield in 1958 when the people were moved out, the Grade School was closed and the remaining buildings were demolished. After the buildings were gone then the levels of the open pit were extended into the area once occupied by the town. Everything that remained, such as concrete foundations, were mined away as if they were common waste. Some of the shovel runners and other mine personnel did their work with regret and sympathy as they assisted in the final demise of buildings that included their old homes and the Grade School where they had spent many years of their lives.

Figure 126 – View From Above Fox's Switch of Remnants of Copperfield In 1958
Photo Courtesy of Kennecott Utah Copper - The Grade School, with the chimney for the furnace, is still standing as are some of the larger walls of several other buildings on Main Street. At least one of the buildings in Jap Camp appears to be undamaged. However, every vestige of Copperfield will soon be gone as the mine expands in this direction and all remaining bits of the town will be mined away. Active mining operations are evident on the upper right side of the photo, on H-Level which was between the Upper Terrace Heights and Dinkeyville.

Figure 127 – The Mining Away of Terrace Heights in 1958
Photo Courtesy of Shirl and Isabell Scroggin – Some walls of the schoolhouse are still standing but all of Terrace Heights is gone except for the concrete retaining wall of the Lower Terrace and scattered chunks of concrete where the Upper Terrace used to be. The shovel and locomotive on the skyline appear eager to mine away the few remains and obliterate all signs of Terrace Heights; they are apparently on G-Level (the level below H-Level) that has been recently extended to where the Upper Terrace used to be. The fire hydrant shed in the upper right corner was the one just above Gonzales's and Miller's house on the Upper Terrace -- it looks like the last untouched outpost of what used to be Terrace Heights and Dinkeyville.

COPPERFIELD RESIDENTS WANT POST OFFICE

On Thursday evening last a mass meeting was held in Copperfield with the idea to petition the Postoffice Department to establish a Postoffice there.

Inspector C. F. Dutton of Denver, Colorado, who has been here the past week investigating the Bingham postoffice, was present. John Knudsen is to be congratulated for his dogged persistency in bringing this matter to the front.

One mail carrier, mounted, delivers mail daily to the people of Copperfield. Dinkeyville and the outlying districts of the U. S. Mines. If the P. O. Department deem it advisable for a postoffice to be established at Copperfield it will remain to be proven in the near future.

The above article was printed in the July 22, 1922 edition of The Press-Bulletin, Bingham Canyon, Utah.

Figure 128 – Princess Theater and Copper King at Main Street and Carr Fork About 1961
From the Gordon Bodily Collection - The Princess Theater at this location was the show house that most of us grew up with but now it has reached the end of its life and will soon be demolished. All the buildings on Main Street above the Copper King have been demolished or covered with waste rock. The Bingham-Copperfield vehicular tunnel entrance is to the left of the photo; it is also being covered over. The company offices up Carr Fork and the access road to the 6190 Yards were still in use when this photo was taken. The last movie at the Princess Theater was shown September 25, 1960.

LAD HAS NARROW ESCAPE IN FRONT OF PRINCESS

Charles, the eight-year-old son of Mr. and Mrs. G. A. Schmitt, had a narrow escape Thursday night of last week just after coming out of the Princess Theatre, when he was knocked down by one horse and run over by three others. Fortunately the boy received no serious injuries.

Just as he came out of the show and was walking along the sidewalk he was struck by a loose horse and knocked into the street, and before he could get up was run over by three other loose horses. There is no restrictions in Bingham against horses running at large in the streets.

The above article was printed in the February 1, 1918 edition of The Press-Bulletin, Bingham Canyon, Utah.

Figure 129 – Lower Portal of Bingham-Copperfield Tunnel Being Covered About 1961
From the Gordon Bodily Collection - A top corner of the vehicular tunnel can be seen near the center of the photo in the midst of the waste rock that is being pushed over it by a bulldozer. The 6190 Yard buildings are on the right edge of the photo with the pit and some East Side levels in the background. The vehicular tunnel was the link between Bingham and Copperfield for nineteen years (1939 to 1958). Kids going to school in Copperton rode the schoolbus through the tunnel twice a day. It was also used frequently as we went to the Saturday serial movies in the Princess Theatre, to the Gemmell Club, and for other less regular needs. All of the mine operations in this photo are shovels loading railcars; haulage trucks will be used starting in 1963.

Figure 130 – Construction of the Yosemite Shops in 1963

Photo Courtesy Of Kennecott Utah Copper - A Level area was excavated into the hillside and then enlarged by dumping waste rock to prepare the site for constructing the Yosemite Shops. Some of the old landmarks can still be identified but many have been covered over by the waste rock. The old road that went just below the Yosemite and Brooklyn mines can be seen on the left side of the photo halfway up. The Yosemite and Brooklyn mines and their dumps were down there near the center of the gulch as was the "shortcut" tunnel between Copperfield and the Yosemite. This was an area where we had many hunting experiences including our dog Tip's worst encounter with a porcupine by that old road. A deer attacked Bert Ivie just below the old road by the low saddle more than halfway up the photo. This huge truck shop is visible on the back cover of this book as a tiny building in comparison to the mammoth open pit of 2003.

Figure 131 – Empty Space Where Copperfield Used To Be – About 1973
From the Gordon Bodily Collection – This photo of the Utah Copper open-pit mine shows almost no trace of Copperfield. The entire community has been mined away so that old-time residents can only speculate as to where, in the air above the pit, their old home used to be. The Yosemite Shops are in the lower, center part of photo; they were constructed in 1963 and truck haulage was also started in 1963. Considerable haulage of waste material by trucks had been accomplished before this photo was taken. Bear Gulch is on the lower left side of the photo. Part of the Telegraph-to-U.S. road is near the left side a third of the way up the photo where the original hillside is in place. The community of Telegraph is under the fill below the U.S. road. The main part of Copperfield has been mined away; it would be up in the air above the mine levels to the left of and just beyond the large area of original hillside above the Yosemite Shops.

MEASURING "PROGRESS" and DATING PHOTOS

There were stages in the development and in the destruction of various features of Upper Bingham, or Copperfield, that can be used to mark the progress of activities and for dating photos. These stages are evident from the early years of underground mining, to the first years of development of the open-cut mine, to the encroachment of the open pit upon Copperfield, to the demise of the town in 1958, and to the on-going development of the mine today. There are thousands of photographs of Bingham and of Copperfield that have preserved much of the hundred and forty-four year history of the area. The year the photos were taken can be approximately determined by the stages of "progress" that are evinced by the presence or absence of recognizable features. Some photos have dates attached to them -- they are especially useful. Some features were especially pertinent in the history of Copperfield and were visible in many photos. These features are highlighted in the listing below. Development of the Utah Copper open-pit mine, construction of the Copperfield School and of the E-Line Bridge, use of steam or electrical power for shovels and locomotives, Utah Copper Company housing developments, construction of the vehicular tunnel, and the model years of automobiles were the very most significant.

From 1863 to 1906 there were dozens of underground mines with their attendant mills, smelters, roadways, ore haulage facilities, mining equipment and buildings. Many of the mines persisted for eighty years or more but the duration of some other features was quite fleeting. Following are some date-markers for the Copperfield area:

- **The first recorded mining location in Utah, the Old Jordan Mine, was located in 1863.**
- Placer gold was discovered in 1864 and mining started in about 1865-67 and essentially ended by about 1902.
- The Holden Tramway, with the use of mules, operated from the early 1890s to 1901.
- **In 1901 Shay engines came into use on the Holden Tramway which became the Copper Belt Railroad in about 1902.**
- From 1902 to 1907 the Copper Belt Railroad used Shay engines to haul ore from underground mines.
- From about 1902 to 1910 the U.S. Mine used an aerial tramway to haul ore from its Old Jordan and Galena Mines in Galena Gulch down to Lower Bingham.
- Prior to 1906 there was no open-cut mining activity on "The Copper Hill" between Carr Fork and Copperfield.

After 1906 the open-cut copper mine and the community of Upper Bingham (Copperfield) with their features provided the best means of dating photos.

- **In 1906 the first open-cut mining operations began with the Boston Con near the top of The Copper Hill and the Utah Copper at the bottom.**
- In about 1907 the East-Line Extension Bridge was built.
- In about 1907 construction was completed on the D&RGW High Line; it serviced the Utah Copper, the U.S. Mines in Galena Gulch and underground mines in Copper Gulch and Highland Boy.
- From 1907 to 1911 the D&RGW High Line hauled all the ore from the open-cut mine to the mill at Magna via the Cuprum Yards with standard-gauge steam engines. It continued to haul a portion of the mine production until 1920.

- In about 1908 the Upper Crossing (H-Level Wooden Bridge) was built.
- **In 1909 the first Upper Bingham (Copperfield) Grade School opened.**
- By about 1911 the virgin hillside between the Boston Con and Utah Copper open-cut operations was all mined out and the two mines merged.
- In 1911 the Bingham and Garfield Railway came into service hauling some of the ore from the B&G Yards above Carr Fork to Magna.
- From 1911 to 1920 both the B&G and the D&RGW High Line were used to haul ore from the open-cut mine to the mill at Magna.
- **In 1912 the E-Line Bridge was built.**
- Greek Camp was built between 1909 and 1915.
- Jap Camp was built between 1909 and 1915.
- Between 1916 and about 1929 the "Upper Crossing" wooden H-Level bridge was torn down and replaced by the concrete overpass.
- **In 1918 the first Upper Bingham (Copperfield) Grade School burned down.**
- **In 1919 the second Upper Bingham (Copperfield) Grade School opened.**
- In 1921 most of the hillside between the east abutment of the E-Line Bridge and H-Level was still native, unmined hillside.
- **In 1921-22 the Utah Copper Company houses were built along Main Street.**
- The U.S. Hotel was built between 1920 and 1926.
- **In 1924 the one-story duplexes of Terrace Heights were built.**
- In 1926 the auditorium/classrooms/library addition to the Grade School was constructed.
- **In 1925 and 1926 the two-story Upper Terrace Heights duplexes were built; the Lower Terrace duplexes were built in 1926 and 1927.**
- In September-October 1926 the shower house was built in Jap Camp.
- The East Line Extension Bridge was demolished between 1929 and 1941.
- The Miner's Merc was torn down in about 1939.
- In 1937 work was started on the vehicular tunnel from Bingham to Copperfield.
- The older wooden houses below the Lower Terrace Heights were demolished after 1927 but before 1934.
- **In 1939 the Bingham to Copperfield vehicular tunnel was completed.**
- In 1940 the roadway from Bingham to Copperfield was severed.
- **In about 1940 the E-Line Bridge was demolished.**
- The ore bins at the south end of Dinkeyville that emptied onto H-Level and the trestle to the bins were in good shape in 1934, the trestle was partially dismantled in 1944 and the remainder of the trestle and also the bins were gone in 1953.
- In 1945 a large silver-colored equipment shed was constructed on the East Side I-J tailtrack.
- In about 1945 a large equipment shed was constructed on J-Level of the West Side.
- In 1953 or shortly before that a new road was constructed above Dinkeyville from the south end of the community to the JK switchback and up to the Saddle.
- **In 1958 the entire community of Copperfield was demolished.**

The bottom levels of the pit can also be used to determine the date of photos. The center of the higher grade "core" of the orebody was about 900 feet west of the canyon bottom. In the early years of the mine the center of the orebody was part way up the West Side of the mine. Mining of the higher grade ore required that the West Side be pushed ever-farther back to the west as the pit was developed and then deepened. Some pertinent information in this regard are:

- In 1926 the center, or higher grade "core" of the orebody had been mined down to the elevation of D or E-Level and the width of the "pit" was very small.
- In 1926 A-Level was the lowest level in the mine.
- In 1936 6240 was the bottom level of the pit.
- In 1939 6190 was a dropcut that reached a third of the way across the pit with 6240 the other third.
- In 1940 6140 was a dropcut from 6190.
- In 1941 6040 was a dropcut from 6090.
- **In 1944 construction was started on the 3975 foot long 6040 tunnel; in 1946 it was opened to ore haulage.**
- In 1946 6090 was the bottom level of the pit bottom except for the 6040 dropcut.
- **In 1952 construction was started on the 7000 foot long 5840 tunnel.**
- In 1953 5840 was the bottom for two-thirds of the pit with 5890 the other third and the 5790 dropcut was just getting started.
- **In 1957 5690 was the bottom for half the pit with 5740 the other half.**
- **In 1959 construction was started on the 17,050 foot long 5490 tunnel.**

The highest levels on the East Side can be used to determine some dates of photos.

- **In 1924 the highest level on the East Side above Copperfield was H-Level**
- In 1927 the highest East Side tailtrack above Copperfield was K-Level.
- **In 1934 the highest East Side level was N-Level.**
- In 1944 the East Side levels were to the top of the ridge (Q-Level)

The equipment used at the Utah Copper mine provides another way to date photos.

- **Beginning in 1906 steam-powered Standard shovels were used.**
- From 1906 to 1907 the Copper Belt Railroad used steam-powered Shay engines to haul ore from the Boston Con and Utah Copper open-cut mines.
- In 1907 the D&RGW High Line started using standard-gauge steam engines to haul all the ore from the open-cut mines.
- **In 1925 electrically powered shovels were first used on The Hill.**
- In 1926 some electrically powered locomotives were first used on The Hill but steam engines were still being used in 1927.
- In July 1963 the Yosemite Truck Shops were under construction.
- **In November 1963 truck haulage was started on the waste levels of the mine.**
- **In late 1987 the in-mine crusher and conveyor system began replacing the train haulage system in the pit. In 1988 the crusher-conveyor system worked fulltime.**

The **model years of automobiles that are in some photos** provide information regarding the date.

FASTER AND BIGGER

Steady improvements were made in the mining equipment and methods as the years went by and the mine grew larger and larger. The result was that production from the mine rapidly increased almost from the first day of production and the threat to the town of Copperfield rapidly escalated. The earliest equipment of the mine included narrow-gauge Shay steam engines on the Copper Belt Railroad and Standard steam shovels, some of which had been used on the Panama Canal.

Figure 132 – Whirly Electric Shovel Loading Haulage Truck In 1976
Photo Courtesy of Kennecott Utah Copper – Truck haulage by Kennecott was first implemented in November 1963 on the waste levels of the mine. After electric shovels (beginning in 1926) and haul trucks (1963) were brought into the mine they were superseded every few years with ever-larger ones. The capacity of the dipper on this #38 P&H shovel is 25 cubic yards; the capacity of the #321 Lectra Haul truck is 150 tons. In 2006 the largest shovel dipper had a 56 cubic yard capacity while the largest trucks had a 360 ton capacity; it would take the largest shovel about four dipper loads to fill the largest truck.

The Shay engines and the Copper Belt Railroad were replaced by larger standard-gauge steam engines on the D&RGW High Line which ran from the Lead Mine area to the Cuprum Yards on the East Side above Bingham. The Cuprum Yards were on an elevation equivalent to A-Level in the open pit. The High Line gave way in 1911 to the B&G Railway which ran from Garfield along the foothills and then up Bingham Canyon over three steel bridges and through four tunnels to the B&G Yards in Carr Fork. The B&G Railway had a 2.5 percent maximum grade so that articulated Mallet steam engines were able to pull sixty-ton cars in forty-car trains. The steam engines being used in the mine were also upgraded as the D&RGW High Line and the B&G Railway equipment was improved. Beginning in 1926 the steam engines in the mine were replaced by GE electric locomotives which were powered by 750-volt overhead lines. The B&G Railroad from the mine to Garfield was replaced by haulage of the ore from the mine to the Copperton Yards and then by the Low Line from Copperton to Garfield on which larger electric locomotives were used; these electric locomotives were later replaced by diesel engines.

The haulage line from the mine to Copperton was upgraded, as the pit grew deeper, by construction of railroad tunnels that came out farther and farther down the canyon. The 6040 tunnel was completed in 1946; it was 3,180 feet long and came out in the central part of the canyon. The 5840 tunnel was constructed in 1952; it was 6,100 feet long and came out in Lower Bingham. The third and last tunnel, the 5490, was constructed in 1959; it was 17,050 feet long and comes out just above Copperton.

Beginning in 1925 the Standard steam shovels were converted to electric power. In 1936 there were twenty-three electric shovels, all with 4 ½ yard dippers. These Standard electric shovels were in turn replaced by the first electric "Whirlies". The early electric Whirlies were in turn replaced by several generations of newer Whirlies with ever larger dippers. Now, in 2006, a shovel with a dipper of 56 cubic yards, more than twelve times the capacity of the 4 ½ yard dipper, is being used.

Diesel-powered haulage trucks replaced the electric locomotives and rail haulage system in the mine -- first on the upper levels beginning in 1963 and later on all the levels. To begin with the trucks hauled the waste rock out to the dumps and a small amount of ore to a reloading station; trains continued to haul the ore from the mine to the Copperton Yards and then to the concentrators at Magna and Arthur. Then more and more of the lower part of the mine was converted to truck haulage.

Ultimately, beginning in 1987, an in-mine crusher and conveyor belt system replaced the trains for hauling the ore out of the mine and all the way to the new Copperton Concentrator just north of the town of Copperton. The crusher was built in the mine so that the ore could be reduced to a size that the conveyor belt could handle. Trucks with a capacity as great as 360 tons now haul the ore to the crusher which reduces the ore to a smaller size which is then fed directly to the conveyor belt. The conveyor belt moves the ore through the 5490 tunnel to Copperton and from there (via the Overland Conveyor) to the new Copperton mill. The 6040 and 5840 tunnels have been abandoned.

The improvements in equipment are continuing today – mostly by steady increases in the size of shovels and trucks and by use of the conveyor system. However, there is a limit to the life of the open-pit mine that can't be continually overcome by improvements in equipment. Rio Tinto, the present owner of the Utah Copper mine, anticipates the existing open pit will be worked out by 2013 but some open-pit mining will continue until 2017.

The next major change in mining technology for the Utah Copper mine is expected to be that of underground mining that is expected to be phased in as the open-pit operation is phased out. Block caving will be employed on a large body of ore averaging 0.7% copper. A higher-grade skarn deposit averaging 1.89% copper will also be mined underground, probably by a different method.

CHAPTER EIGHT

MY PERSONAL HISTORY AND ADVENTURES

My name is Robert Eldon Bray. I was born January 11, 1934 in Salt Lake City, Utah. My father, LaMar Bray, was twenty-five years old and my mother, Grace, was twenty-two when I was born. My older brother, Garland, was three.

Our little family lived in Salt Lake for only another year after I was born. In 1935 my dad was rehired by the Utah Copper Company and we moved into a house that my dad rented from the company at 691-A Terrace Heights in Copperfield, Bingham Canyon, Utah. Our house was half of a two-story duplex; the Alton Ivie family lived in the other half. This house was to be my home for all the years that I was growing up. It wasn't until 1952, the summer after I finished high school, that my parents moved to Copperton.

MY FATHER

My father's full name was Robert LaMar Bray. He was born August 23, 1908 in his Grandfather and Grandmother Glenn's house at 18 Freeman Avenue in Freeman Gulch, Bingham Canyon, Utah. My father was the third of three children born to Clealon and Jennie (Genevra Glenn) Bray. Clealon and Jennie were divorced a few months before my dad was born. Jennie then moved in with her parents and later married and divorced four times so my dad had several stepfathers but no real father of his own while growing up. As a consequence, he grew up as a belligerent teenager who had little respect for any authority figures and constantly fought with other boys who were anywhere near his size. He had little interest in school and made it into the seventh grade before he quit school to work for the Utah Copper Company. His uncle, Joe Delaney (Emily or Elma Glenn's first husband), introduced him to the prizefighting ring as an alternative to street fighting. LaMar gained a reputation in the ring and earned some money besides. He took up smoking and drinking at the early age of about thirteen and was to pay a heavy price for it throughout his life.

My Dad's bad habits were greatly tempered by his marriage to my mother but he wasn't able to quit smoking or drinking until near the last years of his life. He did get into a few scrapes in the saloons in his later years but not nearly as often as he did in his youth.

Figure 133 – Mouth of Freeman Gulch With Streets and Houses About 1958

Photo Courtesy of Shirl and Isabell Scroggin – This is the mouth of Freeman Gulch where it empties into the main Bingham Canyon. The gulch itself extended a mile and a half above the main canyon and was the range for numerous deer and elk. Near the turn of the century there were several small underground lead-silver prospects or mines farther up the canyon. These mines included the Red Wing Extension mine with three tunnels (the Upper, Lower, and Old tunnels) and the Kansas No. 2 tunnel. Placer gold mining was conducted near the mouth of Freeman Gulch. It was reported that the three Freeman boys were killed in a cave-in at the placer diggings they were working 100 yards northwest from the No. 2 Fire Hall at the mouth of Freeman Gulch (the Fire Hall is in the lower left of this photo). All three boys were buried in a single grave near the mouth of Bingham Canyon opposite the Revere Switch of the Rio Grande Railroad. The author's father, LaMar Bray, was born in his grandparent's house at 18 Freeman Avenue in 1908. The house is almost in the center of the photo as indicated by the arrow; it was occupied in later years by Bob and Maxine Carter and children Richard and Susan. This photo was dated by the model years of the automobiles that can be seen.

Pop, as my brother and I called him, always did as well as he could for his family. He kept food on the table and taught us how to hunt and fish and to work. He discouraged us from misbehavior and was supportive of the things we did right. He was never violent or abusive to my mother or to any of us kids. In all our years of growing up my Dad spanked me only once. He never did spank Garland or our much younger brother Larry or sister Afton. The "spanking" that my Dad gave me was the result of Gar and I not coming home from the schoolyard when Mom turned on the porch light as a signal. Finally Pop came after us and we started running for home with him chasing us. Gar was older and faster than me. In fact he was faster than our Dad. Pop was just barely able to take a swat at me as I ran up the hill and that is the only "spanking" he ever gave me.

My father passed away April 5, 1980 at the age of seventy-one. I was fortunate to have been his son and to have lived with and known him for forty-six years. I remember my father as a "character" who indelibly impressed all who came in contact with him. He was, at the age of thirty-three, 5' 4" tall, weighed 140 pounds and had brown hair and blue eyes. He worked most of his life as a locomotive brakeman and engineer at the Utah Copper (later Kennecott) open-pit mine.

It is with some misgivings that I have written the stories about "Our Father's Trials" because readers may think these stories to be representative of his abilities and actions. It is true that his early life left him with some bad habits such as drinking and smoking that led to some comical situations and "scrapes". But he gave us life, did the best he could for us and set many good examples.

MY MOTHER

My mother was Grace Allene Lumpkin. She was born February 12, 1912 in Myton, Duchesne County, Utah. Her father was Webb Long Lumpkin and her mother was Beulah Mae Elledge. My mother was the fifth of eight children born to Webb and Beulah. She was born in the log cabin her father had built when he homesteaded the former Indian land. She lived on the ranch in Myton for the first eleven years of her life. Webb lost the ranch in 1923 and started anew in Salt Lake City. He and his family were just getting on top financially when catastrophe struck. Webb, while returning home from work on October 15, 1925, had a stroke and died almost immediately. Beulah Mae and her children had a very difficult time living without their father. My mother, Grace, lived with her family and attended school in Salt Lake City until she married LaMar Bray on August 19, 1929.

BROTHERS AND SISTER

My brother Garland was three and a half years older than me. He and I and our parents comprised our family until 1946 when our little brother Larry was born at which time I was twelve and Gar was sixteen years old. Larry was with us for two years when he died of an accident. Our little sister Afton was born in 1949 when I was fifteen and Gar was nineteen years old. During my and Gar's growing-up years Larry and Afton weren't yet born or were too small to participate in our teenage activities. Gar was a really good older brother to me. In spite of my adventures while he chased me into the bathroom and down the trail to the Lower Terrace, he never did hit me or hurt me in any way. We were very close companions and shared many hunting adventures together.

Figure 134 – LaMar, Grace, Garland and Eldon Bray in Salt Lake City in 1935
Photo from Grace Bray – This photo was taken in Salt Lake City a few months before our family moved to Copperfield. LaMar was twenty-seven, Grace twenty-three, Garland five and Eldon was one year old.

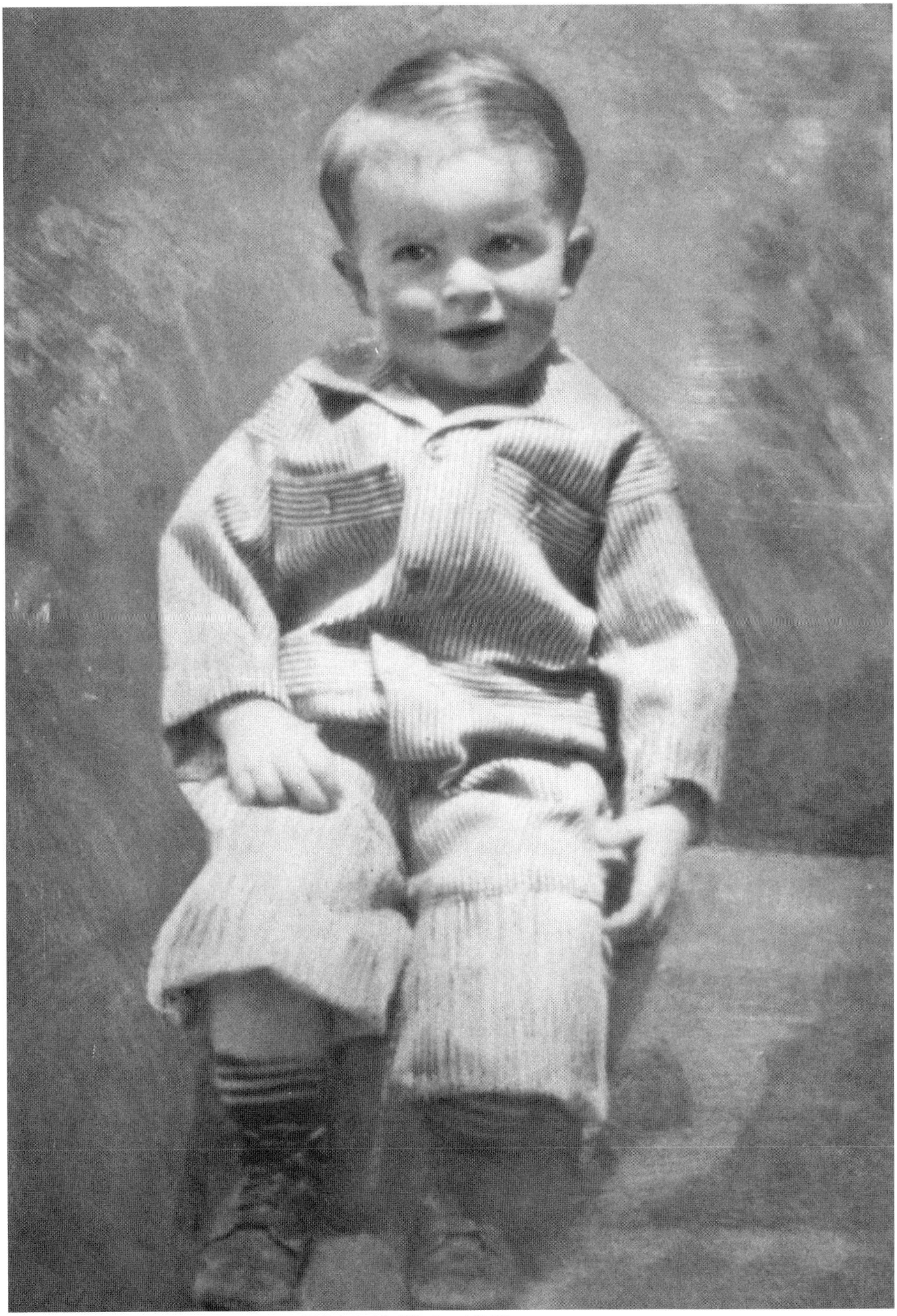

Figure 135 – Lawrence Webb Bray (1946-1948) in 1948
Larry died of an accident just before his second birthday. This tragedy was very difficult for our Mom and Dad. The birth of our sister, Afton, in 1949 helped them to overcome their grief.

Figure 136 – The LaMar and Grace Bray Family in 1967
Photo from Grace Bray – Grace was fifty-five, Afton eighteen, LaMar fifty-nine, Eldon thirty-three and Garland was thirty-seven years of age when this photo was taken. The third of the three sons, Lawrence Webb Bray, died from an accident in 1948 at the age of two years.

Figure 137 – 691-A Terrace Heights, Copperfield About 1927
From the Gordon Bodily Collection - 691-A, the house in which I grew up, is in the center of the photo on the Upper Terrace. It is, of the row of six duplexes, the third from the right, fourth from the left side. The LaMar Bray family lived in this house from 1935 until 1952 when the family moved to Copperton. The Alton Ivie family lived in the right side of this duplex, 691-B. Each side had two bedrooms and a bathroom upstairs, a living-room, a kitchen, and a small utility room on the main floor. The basement had a coal room and a small area for storage and miscellaneous; about half of the basement had only part of the dirt removed and so was about three feet high with a dirt floor. The chimney was on the outside of the house; the heating stove was in the living room and there was an opening for another stovepipe in the bedroom above (See "The Stove Explodes"). The wooden shed on H-Level immediately behind the houses was removed before 1934.

GROWING-UP SCARS

I managed to grow up in Copperfield with very little to show for it in the way of physical damage. The scar on my right cheek is about 1 1/2 inches long and not very noticeable. I received this scar when I was about five years old. There was a large rock outcrop that we called Tommy's Rock just above Dinkeyville and my brother, a couple of neighbor boys, and I went up to climb on it. I can still remember clinging onto the rock and then falling after I lost my balance and grip. Garland led me down the hill to our house with me bawling and bleeding all the way.

A rather serious accident that didn't leave any scars occurred when I was about nine years old. I had some roller skates and used to go skating with other boys in the grade school yard, which had a rather steep slope. We would go to the upper part of the school yard and then coast down toward the fence by the roadway. To stop we would hit into the chain link fence with our stomachs and chests while our arms were outstretched to the side. Unfortunately, this time some workmen had been repairing the fence and the chain link was not in place — only the iron bars were. I was rolling downhill at a pretty good speed and went to hit the fence with my body to stop. The next thing I remembered was lying in a hospital bed with a very painful skull fracture. When I had tried to use the fence to stop, my body had just kept going right through between the upper and lower iron bars and my head had hit the upper bar. Then I had fallen, unconscious, to the roadway below. I was later told that Nick Golish was one of those who came to help and get me to the hospital.

I was in the fourth grade when it became evident that I was near-sighted and required glasses. My mother thought that my need for glasses was the result of my skull fracture only months earlier. I thought glasses were for sissies and I was very upset about having to wear them. However, when the optometrist in a Salt Lake "skyscraper" had me look at people and cars on the street below, first without glasses and then with them, the advantage was immediately apparent and I wore the glasses from that time on but was very self-conscious about them.

The scar on my right hand resulted from a knife stroke by my brother Garland. When I was about ten years old my Dad had gone deer hunting over in Middle Canyon and shot this extremely large 4-point buck up above the old Boy Scout camp in White Pine gulch. He left the deer where he had shot it and the next day got two more men to help him drag the deer down to the road. My mother, Garland, and I also went back to Middle Canyon with him but we waited below at the car while the men went up the canyon to get the deer. Garland and I each had a pocket knife and to pass the time we started chopping down these "Indian corn" plants like lumberjacks felling enormous trees. Garland would make one hack at the plant, then I would make the next and so-on until the plant finally would topple over. At some point, however, with these series of knife chops, we got out of rhythm and Garland took his hack while my hand was in the way. His knife made a deep cut on the back of my right hand that extended from about the center of my hand diagonally across to include my first index or "trigger" finger. The wound bled quite badly and my mother wanted to rush me to the doctor for stitches but my dad was determined to get the deer down to the car first. Luckily no tendons were severed but the finger grew to be smaller than the one on the other hand and it is slightly misshapen.

SHOW HOUSE AND THE ROARING LION

One of the earliest memories I have of growing up in Copperfield was going to the "show house", which was really the Dianna Theater but we always called it the show house. I couldn't have been more than four or five years old at the time so this would have been in 1938. The Dianna Theater closed shortly afterwards. The Princess Theater moved from Highland Boy down to Bingham and it became the show house that we all went to.

I went to the show at the Dianna Theater with my brother and maybe a couple other kids. It was about a half mile walk from our house on the Upper Terrace Heights down to the grade school and then on down the main street to the show house (after the theater moved out then it became the Neon beer joint). It was a whole new experience for me to go to a show. I don't remember the main feature or the cartoon at all but I do remember three things for sure. First, don't sit in the rows too close to the front or it could blind you. Just which rows were safe was indefinite so if a kid dared to sit in the fourth or fifth row he might be safe or maybe not. He might enjoy the show but be stricken blind an hour later. Second, there was a cartoon that was introduced by a brightly-lit screen with a fast-moving, perky, musical score. The kids always hollered and screamed when the commercials were over and the cartoon was ready to start. Third, the feature show for the day opened with a huge, roaring lion. The roaring lion was enough to scare me but nobody else was scared so I hung in there.

After the Princess Theater was moved from Highland Boy down to Bingham (first to a location above the Telephone Office and then to its final location on the west side of the street just below Carr Fork) and as I grew older there were other aspects of the show house that have stuck in my mind. The office for the theater was up front to the right of the stage. There were several times when the owner, Harold Chesler, discovered a rat in his office. He would then get two or three boys to chase the rat around with sticks or brooms and kill it. The boys would then be given free admittance to the show.

The price of admittance was fourteen cents for kids under twelve. Kids over twelve had to pay thirty-five cents. When a kid turned twelve it was to his advantage to pretend to be only eleven. Kids well past twelve would kind of slouch down so that they appeared to be shorter and would reach up to push their fourteen cents in to the cashier. Some kids got away with this for quite a while. One kid got caught in a unique way – he wore his Camp Steiner T-shirt to the show and you had to be over twelve to go to Camp Steiner. The kid paid fourteen cents and was on his way into the show house when Harold spied the T-shirt and hollered "Hey, Camp Steiner! Come back here!" and made him pay the full price.

The weekly serial matinees of superheroes and western cowboys were another big part of our lives. One movie would end at a critical point and we would have to wait until the next Saturday to see whether the hero lived or died.

Getting through the tunnel was another problem associated with anything we did in the main part of Bingham. The tunnel was a mile and a quarter long. Sometimes we walked both ways and other times we tried to bum a ride with a car that had some empty seats. To get to a movie we had to leave home almost an hour early and after a movie we might be so tired out as to dread the walk home which in my case would have been almost two miles. A ride was most welcome but we often had to walk home no matter how tired we were.

TOMMY'S ROCK

When I fell from Tommy's Rock at the age of about five in 1939 I had no idea of how the rock came to have its name. The following decades brought no additional information. Obviously there were a few people who knew its history but I failed to ask the right people. While writing this book and doing research for it I came across a reference of the rock having been named for Tommy Anderson. Sixty-seven years later the answer to a question was found! In June of 2006 Tommy's daughter, Darlene Anderson Householder related to me the story of how the rock received its name.

Tommy's Rock was named after Tommy Anderson. Tommy was born in 1911 and was one of fourteen children. He grew up in Highland Boy and Copperfield. At the early age of about nine, while in church, he had his knuckles rapped so hard with a stick that they bled and so he sought solace by going off alone to cry and think things over. He went to the rock on the hill above Dinkeyville and this became his favorite place to go whenever he had the need to contemplate. From that first time on, when Tommy was missed at home, the family knew they could find him at his favorite rock. So they all began calling it Tommy's Rock. This name was bestowed on it in about 1920. Many people who grew up in Copperfield remember Tommy's Rock but, like me, never knew until recently how it got its name.

SPEARING GRASSHOPPERS

Hunting was a sport that many of us in Copperfield took up at an early age. Most kids start out with flipper crutches, rubber guns or BB guns and a few years later graduate to .22s and then deer rifles. We started out even earlier by hunting grasshoppers with clothespins.

There were lots of grasshoppers up in the hills. They came in a variety of sizes and colors. Some made noise when they flew and others didn't. There were large, light brown ones with bright yellow or even red on the inside of their wings; they made a crackling sound when they flew and they were quite wary and hard to catch. We didn't hunt them very often.

The ones that we hunted most often were about two inches long, heavyset, and were colored medium gray-brown with some black diagonal lines on their backs just in front of the wings. Their bodies, where the wings didn't cover them, were yellowish white. These grasshoppers weren't so wary and didn't fly so far when they were scared up. They were the easiest to catch of all the grasshoppers.

Our weapons were very unique. They were clothespins and match sticks. The clothespins with springs could be taken apart and the spring reversed so that it would hurl sharpened match sticks as if they were spears. One of the notches on the clothespin had to be cut deeper and at an angle to hold one end of the spring. The coil of the spring stood out below the piece of wood and served as the trigger. A "spear" was placed so that the end of the spring, when released, would shoot it forward. The maximum range of the spear was about a foot. We would hunt the grasshoppers and try to shoot them on the ground before they could fly.

Another use for the "guns" was to shoot live matches. When the end of the match that catches fire was placed into the gun and struck by the spring then the match would start to flame as it was shot out of the gun. There weren't any real uses for shooting matches except that it was kind of exciting to see a flaming, smoking match flying through the air. I lost one of my "guns" when the local Sheriff saw me shooting matches and confiscated it.

THE RUNAWAY PUSHCART

H-Level of the Utah Copper Mine ran between the Upper Terrace Heights and Dinkeyville. The track maintenance crew often used a pushcart to move their equipment and supplies along the track. When not using the cart they would lift it off to the side of the track and leave it there until they next needed it.

One evening, after most work crews were gone from the mine, the older kids, including my brother Garland, decided to play with the pushcart. They lifted it onto the tracks and began pushing it along. There was a switch right in Dinkeyville and they turned it to direct the pushcart up the switchback toward I-Level. The track went up at a steep grade around a curve and six or eight kids pushed the cart way up the hill. Then everybody, including myself, piled onto the pushcart and sat down for a thrilling ride. We went zooming down the track, around the curve, and headed for the switch at a great velocity. All the kids were precariously trying to hold onto the flat bed of the pushcart and onto each other for all we were worth. Then we saw a younger kid standing by the switch as if he was ready to throw it. We all started screaming at the kid to get away from the switch but he couldn't tell what we were yelling at him to do. He gaped at us for an instant and then bent over and threw the switch! We were going too fast to even think of jumping off so we all hung on. The pushcart hit the switch, flew off the tracks, and bounced along the ties with all us kids bouncing up and down like marionettes. Everybody managed to stay on and nobody got hurt. The pushcart quickly slowed down and then stopped. We kids all climbed off and moved the pushcart away from the tracks to where it had originally been left by the workmen. We all considered ourselves lucky that nobody had been hurt.

Another incident with a pushcart also occurred on H-Level but this time our father displayed an athletic ability that Gar and the other boys had never before witnessed. Garland and four or five friends again pushed the pushcart from H-Level up around the curve most of the way to I-Level. Then they all jumped onto the pushcart and sat down while it rolled back down the track. It rapidly picked up speed, went around the curve at a fast clip and then, still picking up speed, began the straight downhill stretch toward the switch on H-Level.

Just after rounding the curve Gar looked ahead and saw our Dad standing alongside the tracks. Our Dad began waving his arms and was obviously upset and intent on making the kids stop the pushcart. This was a dangerous game of riding the pushcart at high speed on the railroad tracks. It would have been a disaster if a train had come along and ran into them! Our Dad meant for Gar to get off the pushcart and run home – now! Gar wasn't the brakeman so there was no way he could slow or stop the pushcart. Another boy was kneeling and holding onto the stick for the brakes and he didn't use it. All Gar could do was hang on as best he could to keep from falling off.

The pushcart, with its load of boys, must have been going twenty miles an hour or more as it bore down toward our Dad who started hollering for them to stop. And then – whoosh!! The push cart and the boys, with no braking at all, zoomed right past our Dad who hollered some more and then started running after them as fast as he could. The pushcart had begun to slow down little by little after the track leveled off just before the switch but it was still moving at a considerable speed. Gar had no idea that our Dad could run so fast but he was still being left way behind by the speeding pushcart. Our Dad, however, didn't give up. He continued to chase, with all the speed he could muster, after the pushcart full of boys. He showed unexpected endurance. By the time the pushcart had slowed down to the same speed that our Dad was running the pushcart was eighty yards ahead of him. Our Dad kept running and he was now gaining on it. The pushcart continued to slow down and our Dad continued to gain but the pushcart was still quite a ways ahead. The race went on with the pushcart slowing down and our Dad closing the gap. However, the pushcart was almost stopped before our Dad caught up with it. Our Dad had run at top speed for a couple hundred yards from the switch almost all the way to the ore bins by Carter's house! He was gasping for breath. Gar and the other boys were greatly impressed by the speed and distance that one of their fathers had managed to run but this was no time for congratulations. Our Dad took off his belt and started chasing Garland home, swinging the belt at Gar's rump all the way.

SLOT MACHINE OR LUNCH FOR KIDS

During World War II there was a shortage of labor at the open-pit copper mine so quite a few women were employed in the less physically demanding jobs such as switchtenders and flagmen. Many of these women were mothers whose kids were in school so they had to figure out how to provide breakfast and lunch for the kids. One solution for the grade-school kids in Copperfield was to have them eat cold cereal for breakfast and, for lunch, give each kid a quarter with which to buy a hamburger and soda pop at Frenchy's cafe. However, it was common practice in Frenchy's restaurant as well as the beer joints to let the kids play the slot machines. Eight-year-old kids were well acquainted with the bars, bells, plums, oranges, cherries, and lemons of the slot machines and the payoffs for the various combinations. The owners of the establishment were very willing to let the kids play the slot machines – as long as they lost their money rather than winning. Some kids were so addicted that they put their lunch money into the slot machines and, after losing it all, went without any lunch.

One particularly exciting time for me took place in the International beer joint just below the Lendaris Merc. when I was about ten years old. There were only a couple of customers sitting at the bar and the bartender, or owner, was in his place behind the bar. I, with my friend Keith, had about five nickels with which to play and were putting them into the slot machine and were in the seeming inevitable process of losing them all. The bartender took only passing interest if a kid won three or five nickels. He would show greater interest and perhaps chase him out if a kid won as many as twelve. My luck so far had not been so great as to arouse his interest at all.

But then the first wheel stopped on a bell and the second did the same. When the third reel, with a solid "Kerchunk!" stopped on a winning combination, the machine hesitated for only an instant and then started belching out nickels! Ka-ching! Ka-ching! Ka-ching! The nickels came flooding out. They quickly filled the small coin dispenser on the lower part of the machine and started to fall onto the floor and still they kept coming! Ka-ching! Ka-ching!" I quickly held my hat under the coin dispenser and scooped the nickels into the hat as they came out. The bartender had been alarmed by the first "Ka-ching!" He reacted as if the "Ka-chings" were a fire siren and he was reacting to a life or death emergency. He yelled "Get out of here you kids!" and started running to get around the bar while at the same time yelling at us to stop our gathering of the seemingly endless bonanza that was being coughed out by the machine. Keith was hurriedly picking up the nickels that had fallen on the floor. I kept scooping the nickels from the machine into my hat. The bartender was now rounding the far end of the bar, still yelling at us to get out. Keith and I were in a tense situation! We didn't want to leave any of the money behind but we had to run before the bartender reached us. Just in the nick of time the machine stopped spitting out coins. We grabbed the last of the money and leaped for the door. The bartender was less than twenty feet away when we sprang through the door and ran with our loot! We had won thirty-six nickels!

LAST SHOT WITH A FLIPPER

Hunting squirrels, chipmunks and birds with a flipper crutch was quite challenging. We made our flippers out of the crotch of a maple tree, rubber strips from an old inner tube, and the leather of an old shoe tongue. Smooth rocks about an inch and a quarter in diameter were the chosen ammunition. We would carry a dozen or so rocks in our pocket and shoot at anything of interest such as tin cans and bottles.

The challenge of shooting a live animal was much greater than for an inanimate object and when we spotted a squirrel, chipmunk, or bird then we would immediately begin the stalk. We were lucky to get a decent shot at such game and it was very rare to actually hit the animal. We hunted them for the challenge, not because we really wanted to kill them.

All that changed for me one day when I saw a chipmunk in H-Yards. It was standing up on its hind legs on a rail about twenty feet away and was looking right at me. It was much like the photos that tourists take of the tame little animals in campgrounds begging for food. I automatically took a rock out of my pocket, fitted it to my flipper, pulled back, and let fly. My aim was perfect! The rock flew straight to its target and whopped the chipmunk right square in its chest! I was elated at having made such a good shot. But then I walked over to examine the little animal. It was dead. And it was such a cute, innocent little thing! I had killed it and couldn't bring it back to life. There was no justification for killing it. We didn't eat chipmunks and they didn't do any harm to anybody. Then and there I quit shooting at anything I didn't want to kill.

A DANGEROUS SHORTCUT

There was an inactive mine about one or two miles away on the other side of the mountain. We called it "Bishop's" because that was the nickname of the watchman who lived there. Actually it was the Yosemite mine. There were several other underground mine workings there. One of these mines was a tunnel about six feet wide and seven feet high that went all the way through the mountain. It was supported by timbers but had not been actively maintained for many years. The tunnel had several branches or crosscuts and possibly even some vertical shafts but we didn't know for sure. There was also the possibility of bad air or gas in the mine but again, we didn't know for sure. The other end of the tunnel came out in the upper part of Copperfield (or Dinkeyville) close to Marion Carter's house.

I went through the tunnel a couple of times with older boys when I was about eleven years old. We used flashlights to see with and wore no hard hats or any other safety gear. The older boys had been through before and knew which way to turn whenever we came to an intersecting crosscut. We never had any mishaps but looking back in time it was obviously a very hazardous adventure that could have resulted in serious injury or death.

What was the name of the tunnel next to Carter's house in Dinkeyville that we kids used to go through the mountain and come out on the other side by the Yosemite Mine?

"SAVING" A WILD HOUSECAT

Before I had a .22 of my own (my Dad gave me one when I was eleven) I would go hunting in the hills with my older brother Garland who did have a .22 rifle. One day as we approached the Yosemite mine we scared up a wild housecat that ran and climbed the 8-foot fence that protected the electrical substation. Garland aimed at the cat and was going to shoot it but I begged him not to. I would climb the fence, rescue the cat, and take him home with me. Gar relented and so I went to get it. I climbed up the fence until I was within reach of the cat (which was larger than most housecats). I held onto the fence with my right hand and reached for the cat with my left. The cat immediately bit my hand and wouldn't let go! I jerked my hand away and the cat came with it. He had his teeth sunk in just behind my thumb and his entire weight was suspended there. It was a real shock to have the cat do that to me and it really hurt too. He just hung on by his teeth for what seemed a couple of minutes but was probably two seconds. Finally the cat let go and dropped to the ground. Then it started running across the mine dump toward the tunnel. I hollered at Garland to shoot him but the cat was too fast and got into the tunnel without being hit.

SWIMMING IN THE DRINKING WATER

At least one time, while playing near the water tanks on the hill above Stella's place the older boys decided to go swimming. So they took off their clothes, climbed into one of the tanks and were enjoying themselves with a good swim. And then Stella started screaming. The water tanks were even farther up the hill than the trail from which we would throw rocks at her garages. But somehow Stella had seen what was going on and her voice, with its familiar demands and threats, assailed our ears. "Get out of there you kids or I'll call the police!" Added to her usual cries were the words "That's our drinking water!" I don't recall Stella ever actually calling the police. If she did it wouldn't have done much good because we would have run away faster than the police could run and it would have been too much effort for the police to go after us for such petty misbehaviors.

THE EXPLODING BOTTLE

The underground miners used carbide lamps for their light source and so carbide was very commonly stored or abandoned near the old mines. The U.S. Mine also had a carbide dispenser next to the tracks by Fox's Switch so that the underground miners could fill their lamps before going underground. We kids could generally get some of it from one place or another almost any time. One of the "uses" that we had for it was to blow up glass bottles.

One time several of the older boys were blowing up bottles by the tracks up by Dinkeyville. They had a supply of carbide and several bottles and there was a little six-inch deep pond of water next to the tracks. One boy at a time would take a bottle in his left hand and put a handful of carbide into it. Then, with the lid in his right hand, he would dip the bottle into the pond to fill it about a third full of water, screw the lid on real quickly, stick the bottle down into the mud and run! The bottle would explode in just a few seconds with a loud "Whoomp!" and shower water, mud and glass fragments in all directions so you wanted to be well away from it by then.

One day one of the kids was wearing a new pair of corduroy pants that he'd been given to start school with in just a few days. When his turn came to blow up a bottle, he got the water into the bottle with the carbide and it immediately began to produce acetylene gas. However, when he pushed the bottle down into the mud it didn't stay – it bobbed back up to the surface of the water. He shoved it down again and it immediately bobbed back up. The gas inside the bottle was pretty thick by now. He frantically pushed it back down again as the bottle approached the explosion point. The bottle bobbed to the surface for the third time and then exploded before he could turn to run! Tiny shards of glass as well as a fountain of mud and water shot out in all directions! He was splattered by mud and water and had little bloody spots where the small pieces of glass had cut him. His new corduroy school pants were hit by dozens of the glass shards so that they had little holes all over the front of both pant legs. Luckily enough he wasn't hit in an eye or otherwise badly injured by the flying glass and I never heard what his mother said when she saw him with all the tiny cuts and his new school pants that were now thoroughly muddy and peppered with tiny holes.

ZORRO TACKLED BY GIRL

Every Saturday the Princess Theater would have a serial movie. Zorro was one of the heroes. Besides using a sword, Zorro performed a trademark act of riding a horse. There would be a musical "Tee-ta-dee!" and Zorro would race away on his horse. He was really a gallant hero!

We kids would sometimes play on the hillside between the U.S. ore bins and downtown Copperfield. The hill was quite steep and had quite a few small trees growing on it. There were several trails and one of the main ones went from the highest point on the hill straight down for fifty yards or so. A bunch of us were fooling around playing games. One of the kids was a girl (Shabella) and she was really tough.

A neighbor boy decided to pretend that he was Zorro. He climbed to the top of the little hill and had a towel pinned across his shoulders so that it hung behind him like a cape. Then, just like Zorro, he made a trumpet-like sound of "Tee-ta-dee!" and went charging down the hill like he was on a galloping horse with the cape flowing behind him. He was quite pleased with himself until, near the bottom of the hill, Shabella leaped out from behind a tree and tackled him. She threw him to the ground and totally humiliated the poor kid. To we who witnessed the embarrassing vanquishment of the neighbor boy it was almost as if he really was Zorro and from that time on Zorro was never such a big hero to us.

DYNAMITE FOR KIDS

The town of Copperfield was essentially surrounded by levels of the Utah Copper mine on three sides. A couple of them (H-Level in particular) ran right through the heart of the town. On the East Side of the mine above Dinkeyville there were levels from H clear up to Q which was at the top of the mountain. The Company had one or two watchmen who would patrol the town and were supposed to stop anyone from trespassing onto the Company property. However, the watchmen were pretty easy to evade and I never heard of them catching anybody. We would go onto the switchbacks above Copperfield almost anytime that we wanted to and sometimes onto the main levels between shifts especially if they weren't being worked that day.

Our Dads held almost every job you could think of in the mine. One of them was a powder monkey and he related to his son the proper way to insert blasting caps into the dynamite sticks. There was an iron-walled powder shack on most levels of the mine. The boxes of dynamite were stored inside the shack and the blasting caps, with fuses attached, were stored in a smaller cabinet-like storage area on the outer back side of the shack. These powder shacks would only be left unlocked when a powder crew was working on that level. However, the shack might be half a mile along the level from where the crew was working and this provided ample opportunity for boys to get into the powder shack without being observed.

One day several of the older boys decided to get some dynamite to play with. I went with them up to O-Level where there was an unlocked powder shack. They took half a dozen sticks of dynamite and a dozen blasting caps that already had fuses crimped into them. Garland wrapped the fuses around his neck and stuck the end with the blasting caps down his shirt front. Then we all ran down the hill to the saddle between Copperfield and Lark. Gar was a little bit concerned that he might fall down and set off the caps if he landed on a rock but that didn't slow him down.

We decided that if we were just a bit more cautious we could play with the dynamite and blasting caps as if they were oversized firecrackers. One kid inserted a blasting cap into a stick of dynamite and we all prepared for some kind of special adventure. He lit the fuse and placed the stick of dynamite right in the middle of the open space on the saddle. Then we all ran twenty or thirty yards away and lay on the ground down slope a bit from the flat area so that only our heads were exposed as we awaited the great explosion. I'm not sure what the others expected but the explosion didn't seem very awesome to me. The noise was more than a large firecracker would make and a fair amount of dirt was thrown into the air. But it didn't shake the earth and it didn't make a huge crater in the ground. None of us was hit by flying debris. That was the end of playing with dynamite as far as I was concerned but the older kids still had quite a bit of dynamite and quite a few blasting caps remaining.

Garland and some other kids came upon a 4-inch iron pipe by an abandoned mine below the Lark saddle that was poking up out of the ground at a 30-degree angle. It reminded them of a cannon barrel and they proceeded to use it as such. The blasting caps had 3-foot fuses that would burn at the rate of one minute per foot. They cut the fuse to a 1-foot length and inserted the cap into half a stick of dynamite. Then they lit the fuse and dropped the dynamite down into the pipe. They pushed half a dozen rocks into the pipe on top of the dynamite and ran fifty feet off to the side. When the dynamite went off the rocks went zinging through the air just like from a cannon. The pipe was buried quite deeply in the ground so the explosion didn't burst the pipe. They were able to shoot the "cannon" quite a few times.

At other times some kids cut the fuses to only four inches and threw the half-sticks of dynamite like "hand grenades". Somebody dropped a half-stick charge down the alley behind the school house and it blew out a bunch of windows in the auditorium. The truant officer (Bailey Santisteven) called ten or fifteen kids into his office but never did find out who had done it. Playing with dynamite was obviously dangerous and all the kids who did so didn't escape injury. Elmer Pantalone had a finger blown off by a blasting cap.

If kids today were to steal explosives from a mining operation and set them off like we did they would be branded as dangerous delinquents for life and likely end up in jail!

MICE IN POP'S CAR

On the job Pop and his buddies would often talk fishing. Several times he heard about a red hot fishing spot and we would go there only to get skunked. Then he heard about fishing with mice. The idea was to attach a fish hook to a mouse, put the mouse on a chip of wood, and let it float down the river into the big hole where the huge, hungry fish were waiting for just such a morsel.

I don't know where or how he caught the two small mice but he had them in a fruit jar in the back seat of his car. Garland and I found the mice while the car was parked in the garage. Pop intended to go fishing with them in a couple of days. Garland started playing with the mice, tipping the jar first one way and then the other and watching them as they tried to maintain their balance. Then he decided to unscrew the lid and prod the mice with his finger. When he took the lid off, one of the mice tried to leap out. Gar quickly put his hand over the opening but the mouse was already part way out and squirming for all it was worth. The second mouse immediately joined the first and also started trying to get out. In all the excitement and fast action, both mice got loose in the car. They ran up under the seat cushions. Garland tried to grab them but the springs and stuffing made it impossible. Both mice were irretrievably loose in Pop's car! When Pop found out what had happened he quickly recognized the futility of trying to catch the mice alive and ended up trapping them.

Later on Pop caught another mouse and actually went fishing with it. He attached the hook and put the mouse on a shingle with a string attached to it. He paid out the string and the fishing line together until the shingle was in the fishing hole. Then he jerked the line and pulled the mouse off the shingle and into the water. The mouse swam around for quite a while but no fish ever came after it. Finally the mouse drowned and that was the end of that experiment.

SAVED BY MY GLASSES

I was once saved from very serious injury if not death by my glasses. My older brother, Garland, had a .22 rifle when he was about 14 and I was 10 or 11. I used to go rabbit hunting with him in the hills above Copperfield. One day we went up past the saddle between Dinkeyville and Lark and around onto one of the tailtracks of the open-pit mine.

There were 4-foot boulders at the base of the tailtrack and below that was oak brush — the boulders provided cover, shelter, and sunning spots for cottontail rabbits. We sat on the edge of the level looking down at the bottom of the dump but didn't see any cottontails. So Garland started target practicing. He shot at a little rock in the snow and at a tree stump and then at something else. After three or four shots I asked if I could take a turn. He said "Just a minute" and took another shot. I asked several times but he always wanted to take "One more shot".

Finally, after about ten shots he handed the gun to me. It was a single shot bolt action .22. I put a shell into the chamber, took careful aim, and pulled the trigger. When the cartridge exploded the mechanism holding the firing pin malfunctioned and the firing pin was driven straight back toward my right eye. The pin hit the right lens of my glasses and shattered it; the pin then went flying off in a different direction and we never found it. Luckily, no fragments of glass remained in my eye — perhaps tears washed some out but there was no damage to my eye.

If I hadn't been wearing glasses then the firing pin would surely have been driven right through my eye and probably into my brain. I would have lost an eye and perhaps even been killed! If Garland had taken one more shot before passing the rifle to me then he, who didn't wear glasses, would have been struck by the flying firing pin. That incident stands out as one in which we were both very, very lucky or else somebody was looking after us.

ROCK-THROWING CONTESTS

Stella Klopenstein was an old maid who lived with her brother Alex in a nice house immediately above the Lendaris Merc. Stella also owned some rental garages next to her house. The garages were all joined to each other and had a flat tin roof. A trail ran along the hillside from the Terrace Heights to downtown Copperfield. Stella's garages were a long rock throw below the trail. So every now and then, while taking the shortcut to or from the stores, we would entertain ourselves and test our skills by seeing who could hit Stella's garages the most times with well-thrown rocks. Usually we could hit the garages at least half the time and when we did the tin roof resounded with a "Kerbang!". If we kept it up and hit the garages several times then Stella would run out of her house, look up the hill to see if she could tell who was throwing the rocks, and scream at us to stop or she would call the police. We would duck down so she couldn't see us and no harm came to Stella, her garages, or us boys so far as we could tell.

But that was only half of it. My older brother, Garland, carried the rock-throwing a step further. Years later he told me that Stella also had a tin roof on her house and that he and his friends would throw rocks to hit Stella's house. The result, as you can guess, was that Stella would come running out, look up the hill to see who was throwing the rocks, and scream at the kids to stop or she would call the police. I think she did actually call the police several times and they would drive up to Copperfield just to satisfy her. They never did catch any of the culprits though.

An additional means of harassing Stella was told by a friend who was a couple years older than Garland. The boys would drag a stick along Stella's picket fence so that it would make a "Tat! Tat! Tat!" sound as the stick hit one picket after another. As you can guess, there was an immediate response from Stella who would come running out of her house to stop the kids or try to see who it was and threaten to call the police. I didn't feel sorry for Stella while I was growing up but now, in hearing of things the other boys also did, I do have sympathy for her.

SHREDDED PANTS ON THE PIPELINE

There was a pipeline near the water tanks above Stella's house that went straight down the hill at quite a steep angle. The pipe was about 14 inches in diameter and looked perfect for sliding down. The only problem was that the pipe was quite old, a bit rusty, and not very smooth. One of the kids decided to slide down it and the experiment was disastrous. He climbed onto the pipe, straddled it with his legs, and slid down backwards. Since the pipe wasn't very slippery he wasn't able to go very fast and had to push with his hands to attain any speed at all. Finally he gave up because it wasn't much fun. Then we all looked at him. The whole crotch of his pants was worn clear out by the rough pipe which had acted like a rasp. His pants were ruined and it hadn't been much fun anyhow so nobody slid down the pipe after that.

THE ROLLING, BOUNDING TIRE

There was another experience that I shared with Mr. Katis (though he didn't know it). One day I found a fairly new "knobby" automobile or truck tire that was a bit larger than a regular tire. I started playing with it by rolling it along in front of me wherever I was walking. The tire and I were going down the trail from the roadway between the Upper and Lower Terrace Heights to Bert Ivie's house on the Lower Terrace. With some difficulty I kept the tire under control until I stubbed my toe on a rock and fell forward against the tire. One of the knobs hit me on the nose and bloodied it and the tire started rolling out of control faster and faster. It rolled to the right off the trail and rapidly gained speed as it went down the hill toward the Shay Track. The tire must have been going forty miles an hour when it reached the Shay Track and launched itself in a tremendous bound out into the air. Then it came hurtling down, out of the sky, into the midst of Mr. Katis's vegetable garden. Mr. Katis came rushing out to where the tire had smashed into his garden before it rolled off into Main Street. He looked up the hill in a vain attempt to spy the source of the runaway tire. It all happened very quickly and I was still lying on my stomach where I had fallen. I had only raised my head to watch the tire as it surged down the hill and didn't move when Mr. Katis was looking for its source. Mr. Katis didn't see me and I didn't tell anyone – I was the only person in the world who knew where the rampaging tire had originated.

THE STOVE EXPLODES

The heating stove in our house was in the living room – directly below Mom and Pop's bedroom. The stove pipe went from the stove over to the outside wall and into the chimney that ran up along the outside of the house. In Mom and Pop's bedroom there was a round brass plate that covered the hole for a stove pipe in their bedroom if they wanted to have a stove there.

Pop usually worked overtime on the night or graveyard shift and either slept in late or didn't get home until after Garland and I had gone to school. Mom stayed in bed until after we had left. Each school day morning Gar and I would jump out of bed (it was cold), run downstairs, put the wadded-up newspaper, kindling, and coal into the stove and light the fire. As soon as the fire started I would often climb on top the stove and wait for it to get warm. Sometimes the fire didn't start very well and it would take longer for the stove to get warm and for the room to become halfway comfortable.

Another method of caring for the stove was to "bank" the fire the night before by putting larger pieces of coal off to one side away from the flames. The bigger pieces of coal would burn more slowly or even smolder until morning. In the morning we could often get the fire going again just by stirring the smoldering coals and adding some paper or kindling or smaller chunks of coal.

One morning it was Pop's day off so he came downstairs while Garland and I were getting ready for school. The fire wasn't burning very well so he showed us a miraculous remedy. He got a can of coal oil and poured it onto the fire. The fire immediately roared up higher and hotter than we could have imagined. Wow, the room heated up in no time flat! Gar and I were really impressed.

A few days later we were getting ready to go to school but the fire had been banked the night before and wouldn't restart very well at all. Maybe the paper or the kindling was damp. Garland now knew what to do though – he got a can of coal oil and poured it onto the smoldering fire just like Pop had done. But he put so much oil on that it smothered any flames that might have been there. Instead of roaring to life the fire still just smoldered. We had to get ready for school so we started getting breakfast.

Then – Kaboom! There was a tremendous explosion in the stove! The door to the stove blew open but most of the force went up the chimney.

Mom, who was laying in bed right above the stove, was awakened by the boom of the stove blowing up but also by the brass plate in the wall above her head – it shot across the room as if from a cannon. And then the soot from the chimney began to float down onto her and onto the bed like black snow. There was a lot of it. The whole room was soon covered by a carpet of soot.

Mom was very agitated by the experience. She made Garland stay home from school to vacuum up all the soot in her bedroom.

Almost sixty years later the episode of the exploding stove is remembered as one of the highlights of our growing up in Copperfield. Just the mention of it in our family is sure to bring giggles of amusement and to start another discussion of "Remember when …. "

SNAKE IN THE BOOT

Butterfield Canyon was about a 4-mile hike from Copperfield and we boys used to go there several times per year just to have something to do when we were bored. Butterfield had a small stream of cool, clear water that (as far as we knew) originated from an underground pipe at the old Scout Camp (where the Tooele Fork meets the Left Hand Fork of Butterfield Canyon). Watercress grew in the creek and for a short while there were trout that could sometimes be caught by hand. We often saw water snakes next to the creek and were quick to catch them whenever opportunity allowed. One time in particular I caught a water snake and brought it home as a pet.

Pop was working on the trains either as a brakeman or an engineer. He wore black leather slip-on engineer's boots that were about fourteen inches high. There were no shoe laces or zippers – he just pushed each foot down into them and twisted his foot at the bottom for the final fit. He was working the day shift and went to work especially early because he was part of the coach crew that delivered other workers to their jobs. At the end of each day he also worked an hour later to pick up the other workers from their jobs and bring them to the main pickup points. Before bedtime each night he would set his boots against the wall in the kitchen, all ready to be put on the next morning.

After returning from Butterfield with my water snake I played with it for quite a while but eventually had to go to bed. We had a one-quart 7-Up bottle that looked just the right size so I put the snake in it for the night (without a lid) and placed the bottle right next to Pop's boots.

Real early the next morning we heard Pop squeal with surprise. Gar and I ran downstairs to where Mom and Pop were in the kitchen. Pop was still a bit upset and Mom was laughing. Then we heard the story.

Pop had been getting ready for work. The last thing he did was to put on his boots. He sat down in a chair and pushed and pulled to get the left boot on. No problem. But his right boot wouldn't go all the way on. It felt like he'd left a sock in the bottom so he took it back off and tipped it upside down to shake it out. Instead of a sock he shook out the 12-inch snake! The sudden appearance of the snake startled him and caused the shriek we had heard.

I think Pop's "squeals of surprise" were to some extent his way of calling others to share the incident with him. The snake had crawled away after Pop dumped it on the floor but Gar and I had to look inside the 7-Up bottle to make sure it wasn't still in there where it belonged. Apparently the snake could "climb" much better than I had anticipated and the snake preferred Pop's warm boot over the cold 7-Up bottle.

SINGED EYEBROWS

An experience that I personally had with carbide occurred when Warren Palmer and I were hunting up by the Silver Shield mine above Copperfield. On the mine dump there was a 45-gallon drum about a fourth full of water. Just a little ways off was a shack with a couple pounds of carbide in it. So we did what only seemed natural. We poured the carbide into the drum, stepped back a foot or two and I lit a match. I leaned forward to toss the match into the barrel and before I could get back away it exploded with a loud "Whump!" The explosion luckily was directed mostly upward but it still caught me across the top of my head so that my eyebrows were singed. I was wearing a baseball cap so the hair on top of my head wasn't burned. It took several weeks for my eyebrows to return to normal.

RUNNING FROM THE "LAW"

Most kids who grew up in Copperfield were wary of the company watchman. Parental supervision was generally quite lax so that most kids were free to find their own sources of entertainment and to go where they chose. These activities often conflicted with the responsibilities of the company watchmen. Being chased by "Old Bill Plumm" or "Old Man Gardikas" was part of most kids' lives. The major reason was that the town was surrounded by property that belonged to the Utah Copper or to the U.S. Company. Not only was the town surrounded but there were Company railroad tracks, switches and yards which extended into the town and separated some sections of town from Main Street. To get to Greek Camp or Jap Camp you were supposed to follow the walkway through Fox's Switch but H-Yard was often used as a shortcut. To get to Dinkeyville you had to walk across H-Level. If you strayed from the path when going to these areas then the Utah Copper company watchman would be likely to chase you away. Some houses backed up against company property so that a kid who ventured more than twenty feet from his back door would technically be trespassing. There was limited space for kids to play and to wander. Therefore most kids ventured into areas that were supposed to be off-limits to them. Boys especially were wont to explore the nearby levels and tailtracks of the Utah Copper mine or the ore bins of the U.S. Mine. Numerous buildings were vacated and left standing as the open-pit mine expanded– these became part of the kids' "playground" and another source of friction between the kids and the company watchman.

Bill Plumm retired while I was still quite young but I still remember his name. Mr. Gardikas was the watchman who I spent my growing-up years running from. He chased us away from Fox's Switch, away from H-Yards and away from the coaches that were parked on the sidetrack by Dinkeyville. He chased us away from so many places so many times that I lost count. Running away from the watchman was routine.

Looking back now, from an adult's perspective, maybe those watchmen were actually good, responsible men who cared about the kids. I never heard of Mr. Gardikas catching anybody. All he did was chase them. He obviously knew the identity of most of the kids but he never reported them to the authorities so far as I know. And he never went to our parents to complain of our activities. He just chased us and chased us for years and years while we were growing up.

RIDING THE ENGINE WITH MY DAD

My Dad worked the night shift most of the years he was with the Utah Copper and Kennecott. Most of the time he also worked overtime with the coaches picking up and delivering the men to work before their shift started and bringing them back to the yards after their shift ended. I think the night shift and overtime was, for my Dad, a means of avoiding the temptations of night life in the saloons and of bringing in extra income for his family.

The coach assignment generally called for him to travel the Upper East Side levels when other mine activities and traffic were sparse. This is when he often saw deer, rabbits and other animals. It also gave him the chance to take me with him at times when supervision was lax and few other workers were about. I would climb onto his engine on H-Level and ride with him up and down the East Side switchbacks and across the upper levels for an hour or two. He taught me what the controls were for, about the controller and the brakes, the pantograph and the sidearm. The compressor, which built up the air pressure for brakes on the train, didn't require any care on his part but it was quite noisy. These were enjoyable times for both of us as he did his job with me by his side.

One funny thing about my Dad's working on the Upper East Side happened on the day shift during World War II. Because of the labor shortage the company hired quite a few women to work as "flagmen" or "switch tenders". One of the "flagmen" was our neighbor, Allene Ivie, who was on the J-K flag stand. Ivies had a love for livestock and they usually had some horses and sometimes other animals. One day when there was fresh snow on the ground my Dad and his brakeman were going up the switchback past the J-K flag stand when they saw some fresh deer tracks. Of all places, these deer tracks led right into the outhouse close by the flag stand! Pop quickly put on the brakes and he and his brakeman climbed down from the engine. They cautiously approached the outhouse with the intent of slamming the door shut and trapping the deer inside. If they could lock the deer inside the outhouse then they could come back after work with a gun, shoot the deer, and then pack their venison home. All went according to plan until they were ready to slam the door. It was then that Mrs. Ivie hollered "That's Annie in there! You get away and leave her alone!" Sure enough - rather than a deer it was Annie, Ivie's pet goat, in the outhouse. That was the end of their anticipated extra ration of meat and so, a bit subdued, they climbed back onto the engine and continued up the switchbacks.

Figure 138 - #710 Engine with Coach for Hauling Men to and From Work
From the Gordon Bodily Collection - The engine and the coach or "crummy" have "Utah Copper Company" identification so the photo was taken before Kennecott took over the Utah Copper in 1936. There were two sets of controls for the engine, one on each side so the engineer used the appropriate set of controls for the direction they were going. The brakeman is on the stairs of the coach. The engine has its pantograph up to contact the 750-volt wire that is directly above the engine and attached to the steel tower that is on an extra-long railroad tie. The photo was taken on one of the dumps as attested to by the small steel tower on the tie. On the mine levels the wire is attached to more substantial towers that are farther from the track and the engine would extend its sidearm out to contact the wire. Most engines had a bucket with a dipper for drinking water (see "False Teeth In the Water Bucket") but some crewmen preferred to carry their own water bag.

GOLD AT BEAR GULCH

When I was about fourteen years old my friend Warren and I were hiking above Copperfield near Big Tree. Just below Big Tree (in Bear Gulch) we came upon a placer gold mining operation that was unattended. It must have been late in the afternoon or on a weekend when the miners weren't working. A long wooden sluicebox with riffles was situated a short distance above the bottom of the gulch. They had a gasoline-powered water pump to feed water into the top end of the sluicebox. A hose ran from the water pump down to a small pond that had been bulldozed out of the gulch bottom.

We looked the setup over and decided to give it a try ourselves. Warren, who was a year younger than me, made some adjustments to the engine and started up the water pump. Where or when he learned that much about gasoline engines was a mystery to me but he had a definite knack for such things. With the water running into the sluicebox we next shoveled in some of the dirt that was piled up nearby. After shoveling dirt for half an hour we stopped the pump and searched through the sandy material that had been trapped by the riffles. Gold! I found a small nugget about 3/16 of an inch across. We kept searching and found a dozen or more smaller flakes of gold. The gold wasn't bright and shiny yellow – it was more of a rusty brown – almost the color of tobacco. There was probably a fair bit of iron contained within the gold. But it was gold – the first I had ever seen.

Years later I learned that the first mining claim in Utah had been staked in 1863 only a few miles from this placer operation. The first claims were staked on bedrock lead-silver ore but within a few years placer gold was being mined in several locations in Bingham Canyon. Bear Gulch was one of these sites. The old-timers in the mid 1860's weren't able to find or to recover all of the gold. The operation that Warren and I had come upon in 1948 was at least the second period of placer mining in this area. Another placer operation that I didn't recognize as such in my younger years was less than half a mile above Bear Gulch in Jackass Gulch. There was a very old vertical shaft in the bottom of the gulch that was obviously an attempt to mine placer gold from the sediments near bedrock. Just how successful these mining ventures were I never found out but judging from the amount of gold that Warren and I found I expect they were moderately successful.

TRIP WIRE ACROSS SIDEWALK

Knocking on doors and then hiding or running was a form of entertainment that was done after dark. After the third or fourth false "knock-knock" the man of the house would sometimes become infuriated or maybe just accept the challenge and try to catch us kids. One man on the lower Terrace Heights had a short fuse and would chase us after only the second false alarm. He was in his thirties and in good shape physically so he could run pretty fast. We learned about the effectiveness of a trip-wire by seeing one employed in a movie or maybe somebody told us about it. We decided to try it out on "Leo", who would surely chase us.

Four of us tied a thin rope across the sidewalk about twelve inches above the concrete. Then we knocked on Leo's door and ran and hid. He answered the door and found nobody. We sneaked back and knocked again on the front door. Leo came bursting out of his back door, running at top speed and determined to catch us. We had started running as soon as we knocked the second time so we had a head start but Leo was catching up fast. We ran along the sidewalk as planned. We could hear Leo pounding along behind us, cursing as he ran. We knew exactly where the nearly invisible rope was at and we jumped over it with scarcely a break in our desperate race.

But then there were no more sounds from behind us. No more cursing, no more pounding feet, no outcry of pain. Nothing. Leo had surely hit the rope and been hurled to the concrete sidewalk while running at top speed. We didn't know whether he had been knocked unconscious or maybe even badly injured but we didn't dare go back to see. We never did find out what had happened to Leo after he hit that trip wire. However, we did feel remorse as we came to realize that he could have been badly injured and that it was a very stupid act on our part so we never used the "trip wire" trick again.

STABBING THE FOOTBALL

Some experiences, or pictures, though they may only take a moment, stick in one's mind forever. This was one of them. The kids from down on Main Street in Copperfield were playing touch football in the roadway. They were using Chris Katis's football and were very intent on the game and it was only the first or second down. Then Chris's Dad came out of his house and called Chris to supper. Chris said he'd be right in. The kids played the next down and then one more – they couldn't stop right now! After three or four more plays Chris's Dad came out again and he was furious! He obviously had wanted to start eating a nice warm meal with his family and was very annoyed because Chris's football game had delayed his supper. He ordered Chris to come right that second and Chris did, with the football in hand. His Dad waited for him with a look of impatience and anger. His fuse was obviously very short.

Then came the moment never to be forgotten. His Dad grabbed the football with his left hand and held it tight. Then, with his right, he whipped out a huge butcher knife! He stabbed the ball over and over and slashed it from end to end as if it were an attacking beast that he was fighting to the death. Now its insides were falling out. The group of boys looked on in amazement with their mouths open. When the football was completely shredded and finally dead and Mr. Katis's anger somewhat appeased he chased Chris inside the house to his supper (or worse) and the rest of the boys dispersed.

DELIVERING NEWSPAPERS

Delivering newspapers was a good way to earn spending money. It also was a good way to learn more about the town and to meet people. There were three different newspapers – the Salt Lake Tribune (morning), the Salt Lake Telegram (evening) and the Deseret News (evening). I delivered both the Telegram and the Deseret News at different times. The bundles of newspapers were unloaded by the upper portal of the vehicular tunnel where I would pick them up. There were about sixty newspapers for me to deliver. My route covered the entire town from down by Joe Berger's Combination Bar all the way up Main Street including the U.S. Hotel, then to Jap Camp, clear up to the top of Telegraph, back down to Dinkeyville and finally to the Terrace Heights. It took about an hour and a half to cover the entire route. Sunday morning was more demanding because it was an early-morning paper and was about three times as heavy as the week-day papers.

The U.S. Hotel was kind of special because it was warm and my bag was quickly lightened by about a dozen newspapers. I became very adept at whipping a newspaper down onto the floor so that it slid right through the one-inch opening under the door of an apartment as I walked down the hallway without hardly breaking stride.

The worst part of the newspaper-delivering business was collecting the money. The paper boy had to collect the money each month from each of the customers. Under normal circumstances it was difficult to find each customer at home and so a second or third trip was often necessary. But a few of the customers deliberately put off paying for the paper and made things doubly difficult and disheartening.

Two particularly goofy stories about delivering newspapers actually happened years later down in Copperton but are included here.

The son of my friend Bud was the newspaper boy. The man who delivered the bundles of papers refused to drop them off in front of the boy's house but insisted on leaving them a block away by the store. The man wasn't very consistent on the time that he left the bundles. This necessitated the boy going to the store several times on a stormy day before he at last found that the bundles had been dropped off. And they had been dropped in a puddle of water. This irritated Bud. He telephoned the newspaper company and complained that the newspapers were wet and that a new bundle of dry newspapers should be brought out. The man who delivered the bundles was also quite irritated but said he would bring out some more papers and that the previous bundle "had better be wet!". While waiting for the fresh bundle of papers Bud considered whether the first bundle was really as wet as he'd told the delivery man. So, just to be sure, he stuck them into the running shower! There was no question after that of whether they were really wet or not!

Another incident was when my kids had the paper route down in Copperton. They did a good job day after day and week after week but early one Sunday morning the kids weren't available and I agreed to deliver the papers for them. It had been many years since I had delivered newspapers. There were half a dozen big bundles of papers that had been dumped over by the high school. I looked at them and decided to start out by delivering two bundles at a time. I loaded one heavy bundle of papers into each of the two bags and started out walking to leave a newspaper at each house on the route. After I had delivered about twenty of the papers Kenny Patrick came driving along and gave me the bad news. I had only given them half of a newspaper! I hadn't left the half that contained the funny papers! I looked at the papers still in my bag. Sure enough, each of them was only half of the paper – either the front or the back half. The twenty that I had already left were only half-newspapers and many of them had been taken into the house by the customers. There was no way to be sure whether the customer had received a front or back half except to knock on the door and ask them. Here it was 6:00 o'clock in the morning and many of the ladies were still in their housecoats with frowzy hair and bleary eyes as they came to the door. Most of them were good-natured about my early morning intrusion but a few of them were rather stony-faced because of the unglamorous appearance they had been forced to make.

From "Down Memory Lane, **Business In Early Bingham**" by John J. Creedon
03 Mar 1961(excerpts) Courtesy of Gordon Bodily

Bingham Steam Laundry, later the Royal Laundry, employed many people in Bingham and provided local service for the town. In upper Main there was a couple of Chinese laundries. One of these Chinamen used to do some of our laundry and he would pick up and deliver huge bundles neatly wrapped in newspaper with our name and cost on the paper in Chinese. I never could figure out our name from the price. One time some robbers broke in his place and nearly beat him to death. It was necessary for me to call for our wash and it was a treat to see his laundry. A hot breath of damp air hit me as I opened the door and I could hardly see inside for the steam. The entire room was strung with drying clothes and the smell of incense was heavy. Poor old John was sitting at his ironing board, still recovering from his beating, and I saw my first automatic sprinkling system. He had a pitcher of water handy and he would fill his mouth with water and then delicately blow it over the piece of laundry he was dampening. I had heard about this, but this time I saw it myself. The heat was stifling from the stove where he heated his flat irons. He had several irons on the stove and would change them as they cooled. One handle would fit all the irons.

CHAPTER NINE

MY BROTHER GARLAND'S ADVENTURES

Garland was a very good older brother and we shared many adventures together. But he also did lots of things without me. Following are some of his personal adventures.

AUNT ELMA'S CAT

Aunt Elma and her husband, Jim Thomas, lived across from the upper portal of the vehicular tunnel in Copperfield. Elma was a sister of my Grandmother Genevra Glenn, who passed away when I was three years old. We used to go visit with Uncle Jim and Aunt Elma every few months. Jim had a bulldog named "Cap" and Aunt Elma had a cat whose name I don't recall.

The cat was quite old, had trouble walking, and was suffering from the pain of arthritis or some similar problem. Finally Aunt Elma decided that the cat was suffering too much and should be put to sleep so one day she asked Garland, who was about ten years old at the time, to dispose of her cat. She told him that she would pay him a dollar to do the job and that was big money for a small boy in those days. Gar didn't yet have a .22 of his own but agreed to take care of the cat for her. Aunt Elma provided a gunny sack and coaxed the cat into the bag. They tied it shut and Garland set off to do what had to be done. He took the cat up behind our house to a spot where there were a lot of basketball-sized rocks and laid the cat, still in the bag, on the ground. Then he picked up a big rock and dropped it onto the cat in the bag. At first there was a furious commotion inside the bag but, by the time he dropped the fourth or fifth rock onto the cat, all was quiet. Gar assumed the cat was dead so he untied the string and opened the bag. Zoom!! The cat came running out of the bag like greased lightning and the last Gar saw, the cat was still going down the hill at top speed.

Later that same day Garland went down to Aunt Elma's and found her to be quite disturbed. The cat had returned home and was a bit worse for wear. She decided the cat still had to be put to sleep and she would give Garland a second chance to do the job. This time, however, he had to be sure to do it in a proper fashion. Somehow Aunt Elma managed to get the cat back into the same bag and Gar set off again to dispose of it. He went to our neighbors and got Bob Ivie to shoot the cat with a .22.

Garland had been astounded at the speed and agility of the old cat when it leaped out of the bag and ran down the hill on his first attempt to do away with it. He thought that he had developed a way of rejuvenating old and sickly animals.

GAR SHOOTS HALVERSON'S CAT

Garland was on his way to hunt over the mountain by Bishop's one day. He took the shortcut from the saddle up over the mountain rather than following the trail around the mountain. He was almost at the top of the ridge when he spied a black cat in an oak "tree". The oak bush or tree was only about six feet high. Feral house cats are very deadly predators of rabbits, pheasants, and other small game. This cat was more than half a mile from anybody's house. It had to be a wild house cat. Gar had a policy of shooting feral cats on sight and he did so now. He shot the cat and figured he had accomplished a good deed.

A problem arose later however. Halverson's cat disappeared at about the same time Gar had disposed of the cat on the mountain. Gene Halverson heard of Gar's shooting of the cat so he compared notes with Gar. The color and size of the cat that was shot matched that of the pet house cat that had disappeared. Even though the cat was almost a mile from Halverson's house the conclusion was that Gar had shot their cat! Later on Halverson's cat turned up OK so Gar was let off the hook on that account and nobody else ever mentioned their cat having been lost.

BLASTING CAPS IN JOHNNY'S STOVE?

My brother Garland told me of a stunt that he and some other boys did for entertainment with blasting caps. Many of the houses in Copperfield didn't have a road that ran right to the house. The people had to walk from the nearest roadway to their house and they had to carry their groceries and other items to their house.

One of the biggest jobs was getting enough coal to their house to last through the winter. Most houses had concrete-walled rooms in the basement in which they stored their coal. Most houses also had coal chutes that extended up the hill from a coal bin that was near the house to a roadway that a truck could drive to. The coal chute was really just a large iron pipe about thirty inches in diameter with a concrete box at the top that had an iron lid. The bottom of the coal chute fed into the coal bin which wasn't large enough to hold a year's supply of coal. Usually each family had a big dump truck load of coal delivered just once each year. The coal would be unloaded at the spot that was most convenient for the family – usually this was at the top of the coal chute. The man of the house or kids that he might hire would throw or shovel the coal (depending on the size) into the coal chute and it would slide down to the coal bin from where it would be carried to the window of the basement coal room and thrown inside. Eventually all of the coal would, over the course of the winter, be fed into the stove.

Garland and his buddies thought it would be fun to set off blasting caps in Johnny Barrett's coal chute since it was longer than most and was accessible to the kids. They would light the fuse, throw the blasting cap down the coal chute, and listen at the top end for it to go off. The sound was amplified by the pipe and made a satisfying "Bang!". All seemed to go well with this form of entertainment and Garland didn't think there were any hazards associated with it. But one day Johnny Barrett casually asked Garland whether all the blasting caps had exploded or if there might be some live caps mixed in with his coal. Johnny wanted to make sure that no live blasting caps would end up in the stove in his house. Garland assured him that all the caps had gone off and Johnny never had any explosions in his stove so the incident was forgotten.

GAR THROWN FROM IVIE'S DONKEY

Garland received quite a few scars while he was growing up. Most of them were on his head. He got one of them while trying to ride Ivie's donkey when he was about nine years old. The misadventure started near the orebins just below Carter's house in Dinkeyville. There was an area of bare ground where grass was growing and Ivie's horses (they had about four) were there feeding. Garland was with a small group of boys and when they came upon the horses they decided to take a ride. The smallest "horse" and most tame of the four was a donkey so it was picked as the steed of choice. Gar sneaked up on the donkey from behind and on the uphill side of it. The donkey held still and Gar was able to reach out and touch its back. Then he grabbed it by the mane and jumped onto it. The donkey still didn't move or get upset. When a second kid jumped onto the donkey behind Gar then it started walking and went up onto the road. They started the donkey walking toward the main part of Dinkeyville and were managing to stay on alright. Gar was gripping the donkey's mane with both hands, both boys were hanging on tightly with their legs, and the boy on the rear had his arms wrapped around Gar.

But then three dogs started barking and snapping at the donkey's heels! The donkey abruptly leaped forward and started running! Both boys hung on tighter than ever as they bounced up and down on the donkey but after only a couple of bounces the boy behind Gar slid off the rear of the donkey and fell to the ground. Now Gar was alone, hanging on for dear life and bouncing along on the back of the running donkey while the dogs continued to bark and snarl as they chased them. The donkey, with Garland barely able to hold on, ran along the road into the main part of Dinkeyville. One of the dogs must have actually nipped the donkey because it gave a sudden jerk. That was all it took to send Garland sailing off the donkey toward the concrete steps of one of the houses. He landed short of the steps in the dirt with his hands outstretched to break the fall but his momentum was such that his body kept going and he hit the corner of the steps with his head. His fingers were dug into the dirt and his hands were twisted in such a way that his knuckles were cut by the rocks. He received a one-inch gash on his forehead that bled quite badly. A lady in one of the houses had witnessed the accident. She invited Gar into the house and wiped the dirt and blood away from his hands and head. Then Gar went home. He still has the scars on his knuckles and forehead.

TARZAN FALLS FROM THE "VINE"

Another of Gar's adventures in which he racked up another scar on his head was when he used our clothesline pole like a jungle vine to swing on. He was about eleven years old at this time. There were two identical clothesline poles. Each of them was made from a single piece of iron pipe four inches in diameter. The pipe was bent in two places at ninety degrees so that the ends formed two vertical "legs" with a horizontal support in between. The two legs were cemented into the ground and extended six feet above the ground. The two poles were in the yard by the side of the house – one of them was six feet from the front steps and the other was three feet from the back steps. Clothesline wires were strung between the two poles. The pole by the front steps had concrete beneath it because the window to the coal room was right by it.

The Princess Theater in Bingham was showing Tarzan movies each Saturday. Garland went to these movies and thought that Tarzan, swinging from the vines, was the greatest thing ever. Gar decided that he would emulate Tarzan by swinging from the front clothes line pole. He started out by leaping upward and outward from the lowest of the front steps, grabbing the horizontal section of the clothesline pole and swinging there just like Tarzan. The first step wasn't much of a challenge so he moved up to the second and then the third and the fourth. When he reached the uppermost (fifth) step the pole was below him so that his leap would be downward. This would be a real test of his agility. He leaped for the pole and got hold of it with both hands but his legs and lower body kept moving and swung so high that he lost his grip on the pole. He remembered his hands slipping from the pole but nothing of his movements after that. He must have done a complete somersault because when he came down the right side of his forehead hit the concrete – hard! The next thing he remembered was that of lying on the ground and his head hurt. There were no witnesses so he couldn't say whether he had lost consciousness or not. He went into the house and Mom looked him over. There was a scuff mark on his forehead but no cut on the skin where he had hit. But there was a bump and it kept getting bigger and bigger! Mom put an icepack on the lump and had him lay down on the couch. The lump was about three inches across at its largest and after a couple weeks it decreased to about two inches across but it never did go completely away – he had it for the rest of his life.

ESCAPE THROUGH THE BATHROOM

Garland was 3 1/2 years older than me and when we had arguments that might lead to violence on his part. I often evaded him by such tactics as running into the bathroom which was upstairs and locking the door. I was very proficient at this - I could run up the stairs, dart into the bathroom and lock the door before he, who was only a few feet behind me, could hit the door with his shoulder. After safely locking myself into the bathroom, however, there was the problem of getting back out again. Sometimes I would just wait until he had given up and left. There was another way, however, and Mom was an astounded witness to this.

One day Garland got after me and in a flash I raced up the stairs and locked myself in the bathroom. Mom was in the kitchen washing dishes. The kitchen was immediately below the bathroom and the kitchen window was four feet to the left of the sink and directly below the bathroom window. Garland was especially angry that day and I didn't want to stay in the bathroom for a long while waiting for him to give up and leave. I quietly opened the bathroom window and lowered myself until I was hanging below the window as far as I could reach and then let go. It was a 2-story house and I had about 8 or 9 feet to drop to the ground. I hit the ground with my feet, rolled onto my knees much like a paratrooper, and then took off running. This technique left Garland upstairs waiting outside the bathroom door thinking I was still in there. I might be a mile away before he realized I was gone – it was a very effective way of escape. Mom, however, happened to glance toward the window just as I went flying down past it. She had never before witnessed this escape method and was very startled to see me flash downward past the window at such speed. She hurried outside and looked behind the house but I had already regained my feet and ran out of sight. Nobody was there. It was like she had imagined it all. However, she brought the subject up when we were together later that evening and forbade me to jump out the window anymore.

GAR GETS "CLOTHESLINED"

Garland really was a pretty good big brother. He never did hit me or physically abuse me in any way although he did threaten and chase me fairly often. One time in particular he got the worst of the episode. I was considerably smaller than him which was to my advantage in this instance. He was chasing me down the trail to the far side of the Lower Terrace. Bert and Agnes Ivie lived in the end house and they had a clothesline in their yard. The trail ran under the clothesline. I came running down the trail as fast as I could and ducked only slightly to get under the clothesline and then raced along the sidewalk in front of the houses. After running 100 yards or so along the sidewalk I realized that Garland wasn't chasing me anymore. It was only later that I found what had stopped him.

Garland came racing down the trail in hot pursuit only yards behind me and failed to note the clothesline. It caught him along the forehead while he was running full tilt. His head stopped moving but the lower part of his body maintained its momentum. The impact with the clothesline and the sudden slam to the ground on his back stunned him for several moments. When he came to he gave up all thoughts of chasing me and returned home in a somewhat groggy condition. He was really lucky because it could have been very serious if the clothesline had caught him lower down on the face or neck.

GAR BIT BY A ~~CHIPMUNK~~ BIG DOG

The railroad tracks ran behind our house along H-Level. They essentially separated the Terrace Heights, where we lived, from Dinkeyville which was above us. One day Garland and several of his friends, who were about his age, were playing near the railroad tracks on H-Level behind our house. They saw a chipmunk and chased it. The chipmunk took refuge in the small opening between the angle iron and the rail of the track. The kids could see the chipmunk and there was enough space for them to reach their fingers in to grab the little creature. However, they knew the chipmunk would bite them if they stuck their fingers in.

Garland, as my big brother, ordered me to go home and bring back a pair of gloves. I was less than enthusiastic about being bossed around and very reluctantly started for home at a slow pace. Garland was very impatient so after a couple minutes of waiting for me he took off running for home to get the gloves himself. He took a shortcut down the trail leading to the yard of Mike Callas's house which was on the end of the upper terrace. The steep hillside came right down to Mike's yard and there was a cribbing of railroad ties about four feet high that separated his yard from the hill. The trail went right over the cribbing.

Mike had a big black and white dog that was quite mean. He kept the dog in his yard which was delineated by the cribbing at the far side and a gate he had installed across the sidewalk in front of his house. The dog would often lay in the shade next to the cribbing and considered the yard to be his private domain.

The dog was laying next to the cribbing right below the trail when he heard the heavy pounding of rapidly moving feet coming down the trail. Garland was running down the trail as fast as he could and when he reached the cribbing he didn't break stride at all but jumped off the cribbing into Mike's yard. The dog had been out of his sight up to this point but Garland suddenly was made aware of its presence. As Gar jumped off the cribbing, while he was in midair, the dog, with a snarl, his mouth agape and his sharp teeth ready to snap, leaped for the intruder that hurtled past him.

Gar instinctively jabbed his left hand out at the dog's mouth. The dog instantly snapped his teeth down on Gar's hand. The rapidity of the action and Garland's momentum prevented the dog from getting a good grip, however. Gar hit the ground still running and jerked his hand out of the dog's mouth. He ran across the yard even faster than he had been running down the hill, and scrambled over the gate.

Gar hurried home with his hand bleeding but still very thankful for his escape from Mike's dog. Somehow he had lost interest in trying to catch the chipmunk.

A SACK OF WATER ON THE HEAD

After the Dianna Theater vacated the building in Copperfield then it was made into a beer joint called the "Neon Club". As the Neon it still continued to provide entertainment but of a different fashion. Slot machines and pinball machines could be played by the kids. But there was another, more exciting form of amusement – dropping water bags from the roof down onto drunks below! This was a diversion that I didn't personally participate in but Garland did it a couple of times.

The Neon Club was two stories high and had a flat roof that could be accessed from the hill back behind it. The kids would climb onto the roof and stretch a water hose up onto it. Then they would fill paper bags (the next size larger than lunch sacks) with water. If it was a real big bag they wouldn't fill it all the way but would put only about a gallon into it. They would hold the bag closed with their hands and wait on the roof directly above the door to the Neon, watching for a drunk to emerge.

Most of the men, if they were halfway sober, would come out of the beer joint and walk away so quickly that the kids had little chance to drop a bag onto them. But an inebriated man was very likely to move more slowly and even stand still so that he offered a much better target. Somehow it was more ethical to drop a bag of water onto a drunk rather than a more sober man -- the kids even thought they were performing a public service by sobering up a drunk. But sometimes even a totally sober man might step out of the Neon and pause long enough to qualify for their attention. So when a poor guy, drunk or not, came out slowly enough or if he stopped for a moment then the kids would drop a sack full of water down onto his head.

The bag of water, falling from two stories up, would burst upon impact. The man would receive a double whammy of the sack smacking him on the head or maybe only his shoulder and also of being drenched with water! It was really exciting to see! It would take the victim a couple of seconds to figure out what had happened. Then he would turn his head to look up at the top of the building. The next reaction was for the guy to cuss and shake his fist up at the kids who had ducked back so they couldn't be seen and identified.

The next instinctive reaction for a halfway physically fit man would be to start chasing the smart-alecks (actual words censored but my mother's ears burned). However, the route to the back of the Neon had all kinds of obstacles. The kids had followed a round-about course between buildings farther down the street and then up the hill and along a trail before reaching the back of the Neon where they had climbed six feet up onto the building. The thoroughly angry victim would be faced with fences and houses and bushes that were almost impossible for him to make his way through. The obstacles made it totally impractical to chase the kids so the incensed victim would be reduced to more cussing after which he would stalk away. The kids could sit up on the roof and laugh with impunity. They only did this after dark – it would be harder to get away with in the daytime.

TIPPED OUTHOUSE, ANGRY MAN INSIDE

Halloween was celebrated by the children in Copperfield in various ways. Not everybody had the same enthusiasm for it. Some of us would knock on doors and when the occupant answered we would holler "Halloween comes once a year so give us something while we're here!" and we would be given some candy. But the response might be quite different at other houses. Some residents would turn off all their lights and pretend nobody was home in which case the kids might use soap or even wax to make markings on their windows. Or, as at one house on the Upper Terrace, the kids would be greeted by a bucket of cold water dumped on them from the upper window.

Some of the older kids demanded greater excitement which they found by tipping over outhouses. There were several outhouses in Dinkeyville so this is where they practiced their art. The residents or "owners" of the outhouses were downright hostile toward these pranks and might take violent action if they caught anybody trying to tip theirs over. This necessitated that the boys sneak up in the dark and several of them push together to quickly overturn the outhouse and then run away before they were discovered. The story is told of one boy who, in the midst of the jostling gang, lost his footing and fell into the hole!

Garland participated in one of these forays that proved to be a memorable experience. He with about five other boys crept up to an outhouse late at night and they started pushing with all their might. The outhouse had just started to tilt quite steeply when a loud voice sounded from within! "Who's out there? Stop it! You so-and-so kids!" A man had been using the toilet and he was really mad! The gang of boys all started running as fast as they could but they weren't quite fast enough. The man scrambled out of the privy and threw a baseball-sized rock at the retreating kids. It hit Fino right on the back of his head! It made a big gash which bled profusely but Fino kept running and got away.

Another Halloween activity was cutting clothes lines. The boys (Garland included) carried a pocket knife with which to cut the rope clothes lines. But some of the clothes lines were made of heavy-gauge wire. So the boys innovated – by also carrying a wire cutter! The clothes lines that they cut usually didn't have any clothes on them but there was at least one that did. It would have been especially bothersome to the woman who had done all that washing to have her nice, clean clothes fall to the ground. Cutting clothes lines was more hazardous than tipping outhouses because the clothes lines were closer to the houses. But the boys were quite adept at getting close to the houses because of all the practice they had gained through knocking on doors.

Knocking on doors was another Halloween prank in which the boys would knock on the door and quickly run away so that the occupants would answer the door and (supposedly) be puzzled to find nobody there. Some of the hardier or more determined men (like Nick Nevers) would do their best to catch the boys by chasing them. One really determined man would stealthily wait by his back door and when the boys got onto the porch by his front door he would come running at top speed to catch them. He came close but never did actually catch a single boy.

A UNIQUE "GOVERNOR" ON THE CAR

Gar was going to high school and working on the track gang for the Utah Copper on weekends, holidays, and summer vacations. He was also chasing girls even if he didn't have any spare time.

Gar would borrow Pop's 1937 Pontiac for his dates. Usually he and Lee would go out together but sometimes Gar went alone. Gar was actually a good driver and was very careful with Pop's car. But Pop suspected that Gar was racing it – maybe because that's what he would have done at his age.

One night Gar went out on a date and had to work the next morning. He stayed out later than he'd planned on so it was about 2 o'clock in the morning when he drove up to the lower entrance of the vehicular tunnel. The car seemed to have something wrong with it – coming up Main Street of Bingham it didn't have any pep and he barely made it to the tunnel in second gear and then stopped for the red light. When the light turned green he put the car into low gear and let out on the clutch. The car lurched forward a couple of feet and died! He started the car again, pushed the gas feed to the floor and let out the clutch. Again the car refused to build up any speed and died. The tunnel had a steady six percent grade which wasn't much greater than Main Street of Bingham but he had started up the canyon at thirty miles per hour. Now, starting out from a dead stop, the car just couldn't pull it. There was obviously something wrong with the car and his only alternative was to park it and walk the rest of the way home.

It was a long, exhausting two-mile walk home that night. Gar got a couple hours of sleep and then dragged himself out of bed and went to work. He left a note for Pop about the car and where it was. When he got home after work the car was home – Pop had obviously driven it through the tunnel and up the steeper grade to Terrace Heights. Gar asked Pop how he had gotten the car home and what was wrong with it. Pop was somewhat evasive and, strangely, he didn't seem to be very concerned about the condition of the car. Finally the story came out.

Pop had been so sure Gar was racing the car that he had inserted a homemade "governor" which amounted to a one-inch block of wood with a hole through it for the rod of the gas feed. The gas pedal then could only be pushed down a limited distance. Gar had driven the car the night before with only half the power it should have had! And Pop was responsible for him being forced to walk two miles in the middle of the night when he was so exhausted. Gar was mighty peeved and told Pop so. The "governor" was never used again but it wouldn't have worked anyhow since Gar was now wise to it.

SAVED FROM ELECTROCUTION

Joe Dunn sat still for a moment, wondering what had happened. He had been using the big wrench to loosen the bolts holding the "tower tie" to the rail when an enormous force had suddenly thrown him to the ground. He hadn't lost consciousness and he was still able to move but there was a tingling in his hands and arms. Everybody on the gang stood gawking at the unexpected sight of their Boss's Boss being suddenly flung to the ground for no apparent reason. It took them all a few moments to realize what had happened and then they began to wonder whether Joe might be seriously injured.

Joe, still sitting down, shook his head and looked up at the tower and then at the 750-volt wire 30 feet above him. He could see that the wire had slipped out of its hanger and was in contact with the iron "tower" and then he knew what had happened. He got to his feet and moved his arms and legs to make sure they still worked and brushed off his clothes. Then he looked at Garland and made the statement that Gar has never forgotten. "Boy, it's a good thing you didn't do that – it would have killed you!"

After the track gang had moved the track over closer to the edge of the dump quite a few of the ties had to be straightened, including an extra-long tie that the tower was bolted onto. The tie was also bolted onto the rail and Gar had been using a 30-inch long wrench to loosen the bolts when Joe had come along. Joe was the Track Section Foreman and was quite a character. He was a big man, quite tall, heavyset, and hyperactive. He would get personally involved with the work that the men were doing even when it wasn't necessary, showing them how to tamp ties and perform various other tasks.

When Joe saw Gar loosening the bolts he grabbed the wrench out of Gar's hands and said "Let me show you how to do that!" Joe started loosening the bolts on the outside of the rail that were about 4 feet from the tower. Being hyperactive, Joe would wave his arms around and in this instance the big wrench that he was holding came in contact with the nearby steel tower that was "hot" though nobody knew it – until this moment. He was standing on the ground and made a good enough contact that the 750 volts had given him quite a wallop. Joe was convinced that a smaller man would have been killed and perhaps he was right.

GAR'S APPENDIX BURST

Our family was blessed with good health for the most part. We had our share of measles and mumps but nothing serious until the Fall of 1947 when Garland got sick at the stomach while at football practice. He went home after practice and lay down. The next morning he was still sick and had a stomach ache. He stayed home from school for one day and then two and three. He lay in bed with nausea and stomach pains that wouldn't go away. The doctor and the hospital were down in Bingham and with Pop working we didn't just go to the doctor whenever any of us had a mild complaint. But Gar's pain continued for four days when it became apparent that it was serious. Mom and Pop took him to the doctor who immediately diagnosed it as appendicitis and a very serious case. His appendix had already burst and it was a matter of saving his life! They operated on him and used large amounts of sulfa drugs to fight the infection. Garland was in a bad way and could have died at any time. He lingered on for a week before his condition began to improve. He was very weak for another five days or so. After that he regained his strength more rapidly. He missed almost a month of school. But he did feel better and it seemed that the problem with his appendix was over.

AN ALLERGIC REACTION

Several years after his appendix operation Gar had a nasal infection that wouldn't go away. He had read about penicillin being a wonder drug that could quickly cure infections. So he went to the medical clinic and asked for a penicillin shot. The doctor gave him an injection and Gar returned home. A couple hours after getting home though he got real sick at the stomach and started to swell up. Mom had cooked him some eggs and while sitting at the table he passed out so that his face fell into the dish of eggs. His hands swelled up so big that he couldn't close his fingers to make a fist! They called the doctor and he told them there was nothing to be done about it unless his body went into antiphylactic shock in which case the doctor would give him an injection of adrenaline. So Gar just suffered through the reaction. It took a week before the effects of the reaction were completely gone.

CRIPPLED AND ALONE ON THE TRAIL

Almost a month after having the allergic reaction Gar went with Mom and Pop on a fishing trip to Moon Lake. Pop started fishing on Moon Lake but Garland decided to hike by himself to Bluebell and Spider Lakes which were over the ridge in the Yellowstone drainage and several miles farther north. It would be a long hike – about thirteen miles each way He put a lunch and his fishing gear into a backpack and set off early in the morning. The fishing pole was a two-part type so he put the handle with reel part into the pack and put the longer pole piece partially into the pack so that it stuck up above him like an antenna. He went up the trail along Lake Fork from where it empties into Moon Lake. He followed this trail to the north about four miles. Then he struck off up the draw to the east past East Slide Lake until he reached the level crest of the Lake Fork/Yellowstone Creek divide. From the ridge he climbed down the steep, rocky hillside and it was only an easy quarter mile walk to Bluebell Lake. He laid his pack on the bank next to the lake and immediately started fishing before setting up camp. However, the pole part of the fishing outfit had fallen out of the pack and been lost during the hike. All he had to fish with was the handle with the reel. So he cast the bait out into the lake as best he could and laid the handle with reel down on the bank. He turned back to his pack but was startled by the sound of his line being pulled out from the reel! He grabbed the handle and hauled in an 18-inch trout! So Gar kept fishing and he caught one big fish right after the other until he had seven beautiful 18-inchers. After that the fish quit biting and he didn't catch any more. Then he decided to head back to Moon Lake. He was in reasonably good shape but his knees were a little sore from the morning's long walk.

He climbed back up to the crest of the ridge and then started down the canyon toward Lake Fork creek. It wasn't bad to start with as he went quickly downhill but then his knee joints started aching. They quickly got worse the farther he went! He had to slow way down as he descended the narrow trail. The pain continued to increase. There was nobody else around to help him -- he was completely on his own. Slowly, slowly he continued to hobble his way down the trail. Lifting his feet up off the ground more than a couple inches was especially painful. With every movement it felt like the nerves in his knees were being sandpapered. The exuberant hike up to the lakes had taken only a couple of hours. Now, at his painful, hobbling pace, he was making only a couple of miles in an hour. A six-inch log was too high to step over – he was forced to walk around it. But he still kept going. He was afraid that if he sat down he wouldn't be able to get back up again. He made it to the main Lake Fork trail – it was a bit wider and had a gentler slope but his knees hurt worse than ever. Then he heard a noise behind him. Looking back, he saw a large cow or bull with big horns rushing down the trail toward him. This development was quite worrisome – the animal slowed down but still kept following him and he was afraid it would charge at any moment. But then he got an idea and stepped a few feet off the trail. Sure enough, the cow (or bull) hastened down the trail, passed him, and kept on going. It had just wanted to get past him all the while. He doggedly pushed himself down the trail for another two hours until he at last made it to the Moon Lake campground. It had taken him six hours to make it back even though it was downhill. Mom and Pop could hardly believe their eyes at the bad shape Gar was in when they saw him tottering into camp. He immediately sat down on a bench while Mom desperately tried to do something to ease his pain but there wasn't much anybody could do for him except let him rest. They took off his shoes and gave him water and food but that's about all they could do to help him. Pop fished for another day without much luck while Gar stayed in camp taking it very easy. Then they returned home.

The pain in Gar's knees very slowly subsided and it was over two weeks before he could walk normally without pain. The cause of his ordeal was a renewal or continuation of the allergic reaction to the penicillin shot he had received two weeks earlier for the infection. The basic cause of his allergy might have been the treatment with sulfa drugs he had received for his burst appendix five years earlier. He later learned that, with a reaction such as he had experienced after the penicillin shot, a person's joints are the last body part to fully recover and they can take a very long time.

CHAPTER TEN

OUR FATHER'S TRIALS

A SHOT OF VINEGAR

Our Dad had a drinking problem from way back when he was a teenager. Mom got after him fairly often for drinking too much and this made it difficult for him to enjoy a drink. Garland and I didn't help any either.

One time he was really in a good mood and feeling quite expansive. He went into the back room, got a bottle of whiskey from the refrigerator, and brought it into the kitchen. He had several small wine glasses in the cupboard and he poured a big shot of whiskey into one of the glasses and set it down on the countertop. Then he turned and took the bottle back to the refrigerator. Gar and I were there in the kitchen and a bottle of vinegar was sitting on the countertop close to the glass of whiskey. Gar quickly took another wine glass from the cupboard, filled it with vinegar, and set it on the countertop in place of the glass with whiskey.

Pop came back into the kitchen and with a show of pleasurable anticipation and a happy "Aah!" he tipped his head back and swallowed it all in one big gulp. There was no chance for him to even taste it – the whole big shot of vinegar went down his throat in one split second! But it didn't take long for him to realize his error. He immediately started choking and coughing and got red in the face. The vinegar was all the way down though and there was no taking it back. After he caught his breath he demanded to know what it was that he had swallowed. Rather than becoming angry as some men would have, he seemed rather subdued and disappointed at the bad turn of his anticipated pleasure.

I don't recall our Dad ever being really angry at Gar or I for anything and he certainly never hurt us physically in any way. The thing I couldn't figure out was why he thought the taste of vinegar was so much worse than the taste of whiskey. In my estimation and extremely limited experience I thought it was a tossup as to which tasted the worst.

A SLUG OF BRAKE FLUID

It was about a hundred yards from our house to the garage where the car was parked. Pop would often work on the car and he would sometimes hide a bottle of whiskey in the garage where he could take a nip from time to time without having to walk all the way back to the house or have Mom find out that he had been drinking. He worked on the brakes and the master cylinder several times and used a 7-Up bottle to hold the brake fluid when he bled the air out of the brakes. He also used a 7-Up bottle to hide whiskey in. One time when he meant to take a slug of whiskey he got the wrong bottle and drank a slug of brake fluid! Because of the previous experience with the vinegar he accused Garland and I of switching the 7-Up bottles on him. The funny thing about it was that Gar and I were entirely innocent. We didn't touch his 7-Up bottles. But we never could convince him that he had done it to himself.

DEER MEAT AND THE LION

There were lots of deer and cottontail rabbits in the hills around Bingham – even more than usual because it was a game preserve. Copperfield offered ready access to the hills because there were few or no mine levels above it and there were more pathways leading up into the mountains than in other sections of Bingham. During World War II meat was rationed so that the abundance of deer was especially enticing. Pop and a few of his buddies – especially Nick Golish and Bert Ivie – supplemented their meat rations by poaching deer. The routine was for two or three men to slip out of town with .22 rifles, several flour sacks, a flashlight, and a lunch. They would kill a deer with their .22's, cut it up (often by deboning the entire deer), and carry the meat home in the flour sacks over their shoulders. They usually came home in the dark so that nobody would see them with the contraband.

There were four men on this one particular hunt. They left Copperfield in the afternoon, went up past Big Tree and over into the Queen Mine area of Butterfield Canyon – a distance of about four miles. They were successful in killing a deer and then they spent an hour or so cutting it up and loading the meat into the flour sacks. It was dark before they were ready to start for home. The shortest way back was to follow a water pipeline straight up the hillside to the ridge separating Bingham Canyon from Butterfield Canyon. The oak brush had been cut down for three feet each side of the pipeline so the trail was fairly clear but the oak brush was quite thick on both sides of the trail. Each man carried one or two bags of meat. The man in front also carried a flashlight and the other three walked in single file behind him. The fourth man on this venture wasn't one of their regular hunting companions and he continually dragged behind the others.

Then the man in front saw some eyes reflected in the light. They were bigger and wider spaced than a bobcat – it was a mountain lion! The lion had obviously been attracted by the smell of blood from the deer. All four of them saw the lion's eyes and then they disappeared. They continued on and then they saw the eyes off to their left. And then to their right. Somebody said that a lion would usually pick off the guy in the rear. The guy who always dragged behind was quickly right up there in the middle of the group. They considered shooting the lion but they only had .22's and weren't sure what it would do if they only wounded him. They decided not to shoot the lion but to stay close together and continue climbing the hill and get home as quickly as they could. The lion circled them several times as they climbed up the pipeline; they saw the eyes first on one side of them and then on the other side clear to the top of the ridge. They were a very nervous, even scared group of men all the while until they got home that night.

A DEER ATTACKS BERT

Pop had two special buddies who he went hunting with – Nick Golish and Bert Ivic. All three of them were good hunters and they usually brought home some meat whenever they went hunting. One time Pop and Bert went deer hunting by the spring about half a mile the other side of Bishop's. The water for the "spring" actually issued out of the adit of an old underground mine -- it ran across the nearly level surface at the top of the mine dump and then down the steep outer slope of the dump and into the gulch below. Deer used the spring as a watering hole so it was a good hunting spot.

The two hunters (poachers) stationed themselves near the top of the dump and waited for a deer to come to water. After an hour or so a two-point buck approached the dump. They shot and mortally wounded it. The deer ran down the dump and fell near the bottom where the material from the mine dump covered the lower part of some maple trees. The trees were growing up through the dump material. Bert laid his rifle on the ground and hurried down the dump to the dead or dying deer while Pop stayed up at the top of the dump so that he could shoot the deer again if it got up and tried to get away.

Bert took out his knife just as he reached the deer but it suddenly jumped up and lunged for him! Bert backed away and the deer lunged again. Bert got behind a six-inch tree trunk and the deer still kept trying to impale him with its horns. As soon as the deer made its first lunge at him Bert had started hollering for Pop to shoot it. But the deer and Bert were so close together and moving around so much that Pop didn't dare shoot. The deer finally pushed its head against the tree so that one horn was on each side. Bert grabbed hold of one of the deer's horns with his left hand and held its head so that it couldn't move. Then, with the knife in his right hand, he reached around the other side of the tree and cut the deer's throat.

After the action was all over Bert was still pretty excited and demanded to know why LaMar hadn't shot the deer again. Pop told him that he couldn't get a good clean shot and was afraid of hitting him. This explanation finally satisfied Bert and later on they were able to laugh about the experience.

ONE HUNTER PARTIES WHILE OTHER SUFFERS

One summer day Nick and Pop went hunting deer up in the U.S. (Galena Gulch). They were successful in getting a deer and decided that Nick would walk back to Copperfield and return with an automobile while Pop would clean the deer out, pull it down almost to the road and then wait for Nick on the roadway. Evening was approaching so it would probably be dark before Nick would make it back with his car.

All started out well but Emil Bodmer, who was a part-time game warden, threw a monkey wrench into their plans. Pop made it down to the road and hid his rifle in the brush off to the side. He was prepared to wait there for Nick but Emil's car suddenly came around the bend and pulled up next to him. Emil wanted to know what LaMar was doing up there and Pop gave him some cock and bull story that halfway satisfied Emil. Emil then invited LaMar to get into the car for a little ride. Pop got in the car and Emil drove them farther up the road, turned around, and came back down past where he had picked Pop up. Emil pulled the car over and stopped on the edge of the road where the hill dropped sharply away. Then Emil pulled out a bottle of whiskey and turned on the radio. The two of them swigged whiskey and talked and laughed while the radio played.

Nick reached the road and was hurrying along it toward Copperfield when he heard Emil's car coming around a bend. Nick jumped off the road onto the steep, rocky hillside and lay there waiting for Emil's car to go by. The car went on by so Nick climbed back up to the road and continued walking toward Copperfield. But a few minutes later he heard the car coming back again. So Nick jumped back down the hill again. Instead of the car continuing on past him, however, it pulled over and stopped right above him. Nick could hear the radio playing and the two guys inside talking and laughing. Nick expected that it was the game warden and so he was afraid to move but it was extremely uncomfortable to lay there on the rocky hillside. He thought the car would never leave but it finally did.

Emil drove Pop down to Copperfield and Nick hurriedly walked the rest of the way down. Then they compared their stories and found that, while Pop was partying with Emil, Nick had been laying on the rocks afraid to move. Nick and LaMar went back up to the U.S. the next day to retrieve LaMar's rifle and to get the deer they had shot. Nick and Pop were never sure what Emil would have done if he'd have outright caught them poaching deer but they didn't want to take any chances. They managed to get away without ever getting caught for all the years we lived in Copperfield.

THREE ARMED MEN AFRAID OF A BOBCAT

My Dad, Nick Golish and I turned down a "sure" opportunity to bag a bobcat when we were hunting cottontails one day. We had walked from Copperfield up to the Copperfield-Lark saddle and then dropped down Copper Gulch another half mile to the old Dalton and Lark mine. The old mine dumps were in the center of the gulch; the largest was from the Dalton and Lark mine of which there was a large shaft. However, there were several much smaller adits that went into the hillside for only a few hundred feet – they had far smaller dumps next to their entrances. This was a good place for cottontails -- there were several rock outcrops nearby for the rabbits to burrow under and there was plenty of oak brush and sagebrush for feed and cover.

The three of us split up to investigate promising areas sixty or eighty yards apart. For hunting weapons Nick had a shotgun; Pop and I both had .22s. Nick was walking across the large, open dump when a bobcat, apparently scared up by Pop, ran at top speed across the dump about forty yards in front of him. He was so startled by the bobcat racing broadside right past him that he hollered, raised his shotgun and fired all at the same time. However, the bobcat was running so fast and it was so unexpected that he either missed him completely or at best peppered him with a few #6 pellets. The bobcat didn't even break stride – it ran straight for the entrance of a very shallow mine and disappeared inside.

Pop and I heard Nick holler and shoot. We hurried toward him and he excitedly told us what had happened and we saw the cat's tracks in the snow. There was no doubt that the bobcat was still inside the mine which couldn't be very deep because the dump was quite small. A guesstimate was that the mine only went in less than a hundred yards. But it was dark in there! Within twenty yards of the entrance it was hard to see anything. There could be some small "dogholes" off the main drift – maybe even going upward for twenty or thirty feet. Maybe there were some sharp turns in the drift. I could imagine the snarling, spitting bobcat leaping onto me in the dark from some hidden crevice and scratching and biting me before I had a chance to react. Or, if we cornered him, he might attack us from the front! There were several frightening possibilities of what the bobcat might do if we didn't see him inside that dark, threatening mine and shoot him before he sprang on us. Nick and my Dad undoubtedly had the same thoughts as I did about where the bobcat might be and what it might do if cornered.

We didn't have a flashlight with us but we might have made a torch and used it for a light. Or one of us could have gone all the way back home and returned with some flashlights. I was only about fifteen years old and so deferred to the decision of Nick and my Dad. We tried to look into the mine from the entrance without going in but couldn't see anything. Nick and my Dad, with very little debate both turned and walked away. I followed them. Was it fear or common sense that kept us from chasing the bobcat? Or maybe it just wasn't worth the extra effort. But if we'd have pursued the bobcat into the mine there very likely would have been an exciting tale to tell!

THE FLAMING ASHTRAY

One year my brother Garland and I decided to reload our own rifle shells. It was Gar's idea and I joined him in this enterprise. This was about 1953. By then Pop had moved our family from way up in the canyon down to Copperton which is in the valley at the mouth of Bingham Canyon. We set up shop in the furnace room in the basement. Pop had a workbench down there that was extremely sturdy and heavy. It was about three feet wide and six feet long with a frame of 4x4 inch boards and a top of 2x4s.

We bolted the reloading press onto the workbench and were soon in business. Garland had a .300 Savage rifle and I had a .270 Winchester. We bought the necessary dies and supplies and began reloading our old casings. The powder measure and scales worked OK but now and then, for one reason or another we spilled a bit of powder and maybe even damaged a primer. Pop had a whisk broom and dust pan that we used to clean off the bench top whenever it got too dusty. For lack of a better place we dumped everything that we brushed off the bench into an old ashtray that was on the bench. After several weeks of reloading shells we had dumped quite a bit of dust and spilled powder into the old ashtray.

One day when we were all upstairs, except for Pop, we heard a frightened squall from the basement. Gar and I headed for the stairs just as Pop came running up. He was beginning to settle down after having experienced a great shock. He had been doing a minor repair job in the basement and smoking a cigarette as he did so. When he finished the cigarette he reached over to douse it in the old ashtray. Instantly, with no warning, flames shot a foot into the air and a primer exploded! Pop jumped back and let out a squeal as he did so. None of us had anticipated such an accident. Garland and I must have spilled quite a bit more powder than we realized.

NO BRAKES, SPEEDING BACKWARD

When Garland started high school our father had a 1937 Pontiac sedan. Gar took an auto mechanics course in school and worked on the Pontiac as part of his class. One of the first things he did was to give the car a dark blue paint job. This entailed hours of sanding and other preparation and then the application of the paint. The Pontiac looked much better after the paint job. Then the brakes started acting up so Gar worked them over — this included the installation of a new master cylinder kit. A master cylinder kit included some parts for one particular make of car and other equivalent parts for another make so that there were always some extra parts left over. After Gar completed the brake job and brought the Pontiac home, Pop found the left-over parts from the master cylinder kit and thought for sure that Garland had left some out. So he decided he'd have to do the job over again. He took the master cylinder apart and put it all back together again - whether he made use of some of the leftover parts and whether he bled the air out of the brake lines wasn't clear. But after re-repairing the master cylinder, he decided to take the car for a test drive.

We lived on the Terrace Heights from which there was a fairly level road over to the main road a quarter of a mile away. The main road, starting from the 6% grade of the vehicular tunnel, immediately steepened at the tunnel's mouth so that going up the canyon required shifting down to second gear. Three-fourths of a mile farther up the canyon, by the tailtrack of Fox's Switch, the road steepened much more so that for half a mile only cars with powerful engines could go up in second gear — less powerful cars had to shift down into low. The sideroad from Terrace Heights intersected the main road from the left about halfway up this steepest stretch. A hundred yards above the Terrace Heights road another road from Jap Camp and Greek Camp intersected the main road from the right and immediately above that the main road went through the one-lane concrete H-Level overpass. Just above the overpass the road from Dinkeyville came in from the left and another hundred yards above that the main road finally leveled off a bit as it entered the tiny community of Telegraph.

For his test drive Pop went roaring across the fairly level Terrace Heights road and started up the steep part of the main road in second gear. He made it up past the Jap Camp road, through the underpass, and almost to the Dinkeyville road before he had to shift down into low gear. The engine died as he was trying to get the Pontiac from second into low gear so that it was in neutral at this point. Upon all loss of forward motion it started rolling backward down the steep grade. Pop jammed on the brakes but the pedal went all the way to the floor without any effect. The car rapidly picked up speed going backwards down the steepest part of the main canyon. If left unchecked it might have attained a speed of forty or fifty miles an hour and it might have gone off the road and fallen down the 50-foot drop into the Fox's Switch tailtrack! If it didn't fall into the tailtrack it would still take a lot of luck and skill while rolling backwards at forty miles an hour to keep the car on the main road and then to turn safely onto the Jap Camp or Terrace Heights roads. The car had to be stopped somehow and the situation was getting more desperate by the second! He pulled on the emergency brake – but it failed to slow the vehicle significantly and now the car entered the concrete underpass, still gaining speed. Pop decided this was his only chance to check the rapidly increasing speed of the car. He turned the freshly-painted Pontiac into one side of the concrete wall and then into the other, scraping it the length of the body and fenders on both sides. It worked well enough that the speed was greatly reduced when he came out the bottom end of the overpass and he was able to turn onto the Jap Camp road and get stopped.

It was a thoroughly chastened but also relieved Pop who slowly and carefully drove the Pontiac frontwards, in low gear, onto the main canyon road and, before it gained too much speed, turned onto the Terrace Heights road and limped back home. The car that Garland had so carefully painted and repaired the brakes on was now deeply gouged and scraped along all four fenders and both sides and had no brakes at all.

POP SWALLOWS A TOAD

Mom and Pop lived at 115 West Park Street in Copperton while Eldon and Zona and their children lived at 101 East Park Street - just a couple of blocks away. Mom and Pop were happy to have the grandchildren come visit them which they did almost every day.

Eldon worked in the Geology Department for Kennecott and one day Mom telephoned and said that he should stop by on his way home from work - that something very funny had just happened. Eldon stopped by the house and Mom proceeded to tell him about the extraordinary experience.

Ernie, who was about 6 years old, had caught a very tiny toad while on a fishing trip a few days earlier. The toad was about an inch long. He took it up to show his Grandpa and Grandma. Pop had been duly impressed by the tiny little toad and Ernie left to go home. But he left the toad in a water glass on the sink in Mom and Pop's kitchen. Later in the day Pop had taken a dose of Metamucil and went to wash it down with a glass of water. He picked up the glass that Ernie had left his toad in, and without looking into the glass, he filled it part way with water and quickly drank it, toad and all.

The toad only went halfway down and then it started kicking and jumping in Pop's throat. He couldn't imagine what the problem was but something was obviously very wrong. He started choking and started running for the bathroom but only made it halfway before he fell to the floor on his hands and knees. Mom saw that he was choking and having trouble breathing so she ran over and started pounding him on the back to dislodge whatever was causing him to choke. At this point Pop coughed and out popped the tiny little toad and it started hopping across the floor. Pop, still on his hands and knees, was absolutely flabbergasted and so was Mom. Neither of them could believe their eyes but there it was — the little toad hopping away.

As Mom was recounting the story, Pop was listening to her but not with any enthusiasm. Suddenly he started getting sick and ran again for the bathroom. He said it wouldn't have been nearly so bad if it had been a frog that he had swallowed but the thought of it being a toad was just so repulsive that it made him sick all over again.

POP'S LAST FIGHT

Pop used to go to the beer joint fairly often (to Joe Berger's "Combination Bar" in Copperfield and to the "Moonlight Beer Gardens" after we moved to Copperton). It was nearly impossible for him to always avoid disagreements with other patrons. Usually it was the other guy who started the encounter but this time was unusual because Pop more or less started it and it was with Marv Jensen, the manager of several professional fighters including Gene Fullmer, the world middleweight champion.

Marv was running for County Commissioner and he came into the Moonlight with a friend to have something to drink and maybe pick up some votes. During the discussion Pop, who disliked all officials as a matter of principal, made a remark to Marv about "crooked politicians". Marv, who was sitting on a bar stool, responded by telling Pop "Them's fighting words!" So Pop started fighting. He smacked Marv and knocked him off the stool onto the floor. Now Marv was far bigger than Pop, certainly knew a lot more about fighting, and was considerably younger. But here he was, running for county office and he had just been knocked off his stool by a much smaller, older man in a beer joint. It was quite a predicament for an aspiring County Commissioner. If he were to get up and retaliate it would make quite a story for the newspapers. Pop was obviously prepared to keep fighting and he very likely would have got in some more punches and it could have been a bloody brawl. Marv chose the wiser alternative – he got up off the floor and, with his buddy, left the saloon.

DOCTOR OR UNDERTAKER

Pop's health started failing in his later years. Emphysema, heart, prostate, ulcers – it was quite depressing for a man who had always prided himself on his fighting abilities and physical strength. However, he had no choice but to go to the doctor with his health problems. One trip to the doctor was especially memorable for the exchange between the two.

The doctor had finished examining Pop and was telling him what was wrong, what he had to quit, what he had to do, etc. Finally Pop told the doctor that he would just go see Stevenson. The doctor said "Stevenson! Stevenson! I don't know any doctor named Stevenson!" Then Pop told him that Stevenson was the undertaker.

Boulder Injury Hurts Take Bingham Woman's Life

Injured last Sunday when a 250-pound boulder hurtled down a mountainside and into her home, fracturing her leg, Mrs. Dorothy Sayatovic, 28, Bingham, died Thursday at 6 p.m. in a Salt Lake hospital.

An attending physician said exact cause of death probably could not be determined without an autopsy, but said it is almost certain an embolism took the young woman's life.

This would mean that the marrow from her shattered leg bone had entered a blood vessel and stopped circulation, he said.

Two Leg Fractures

Admitted to the hospital Sunday night, Mrs. Sayatovic was reported as suffering from two leg fractures, a lacerated face and shock.

When the boulder shattered a second floor double casement window in the Sayatovic home, Mrs. Sayatovic and her husband, Pete, were seated on a davenette directly opposite the window.

Mr. Sayatovic suffered a leg laceration, but he was not hospitalized. He was given emergency treatment in Bingham. The couple's daughter, Jo Ann who also was in the room, was not injured.

According to witnesses, several boys were chasing dogs and deer on the hill near the couple's apartment in Bingham, and it was believed their activity may have loosed the rock.

Officers who investigated the mishap were Bingham police officer Elmer Pantalone and Salt Lake county deputy sheriff R. J. Contratto.

Born in Eureka

Mrs. Sayatovic was born in Eureka, Juab county, and was a graduate of Tintic high school. She was married to Pete Sayatovic about six years ago. She was a member of the Church of Jesus Christ of Latter-day Saints.

Surviving besides her husband are a daughter, Jo Ann Sayatovic, her stepfather and mother, Mr. and Mrs. William Kitt, Long Beach, Cal., and a brother and half brother, Jack Ivie and William Kitt Jr., Long Beach.

The above article was printed in a January 1949 edition of The Salt Lake Tribune and was provided by Lorraine Clinton Beckstrand.

CHAPTER ELEVEN

HUNTING

Hunting has been a big part of my life ever since the early age of ten. Why ten? Because that is when my older brother, Garland, first got a single shot .22 rifle and started hunting. I accompanied him most of the time. I was the little brother who tagged along and sometimes was given the opportunity to shoot the gun myself. We hunted cottontails in the hills between Copperfield and Lark most of the time. When I was eleven years old and Garland was fourteen, our Dad gave us each a new bolt action .22 with a 5-shot clip. My rifle was the same as Gar's but mine had a longer barrel. It was a bit difficult to carry and to hold steady but I managed and eventually grew into it. I quickly graduated into hunting with other friends or even by myself but Garland and I still hunted together most of the time for several years. We would walk from Copperfield up over the saddle, down the mountain halfway to Lark, then hunt northward in the foothills to Horseshoe Bend, along the abandoned railroad bed almost to Copperton, then up to The Big Rocks on the top of the ridge, along either the ridge or an upper level of the mine until we were above Copperfield, and then drop down the mountain to our house. That was a long, long way. We walked almost steadily, with numerous side excursions to promising gulleys, areas of thicker brush, and rock outcrops from six in the morning until eight at night. Most of these hunting trips were with our friends Lee Halverson and George Kubota.

After Gar was eighteen he worked for the Utah Copper on weekends and went to school during the week. The only days he had free to go hunting were Thanksgiving, Christmas and New Year's. Lee was working fulltime at the mine and also had these holidays off. We usually hunted on at least two of these days. One of our favorite hunting spots was immediately north of the WyUta at the Wyoming-Utah border up by Evanston. There were a lot of cottontails and quite a few white hares and even a few pygmy rabbits in this area.

Over the years I believe we became experts at hunting cottontails. We knew when they would be out (especially the second good day after a heavy storm) and early in the morning or late in the afternoon. We knew where to find them (brush and rock outcrops) and we knew how to hunt them – spot them at a distance of about thirty yards or, if they started running, wait for them to stop before shooting. Our weapon of choice was .22 rifles – the shells were less expensive, they had a greater range, and they spoiled less meat than a shotgun would have.

White hares called for a different strategy. They would be in the thicker brush and wouldn't stop once they started running. Lee had the best method for shooting them – bam, bam, bam, with a semiautomatic .22 as they were running. Gar and I with our bolt action rifles didn't do as well at getting white hares. Bud Patrick did well with a shotgun.

OUR DOG "TIP"

I say "our" dog because Garland might have been Tip's first real owner but I spent an awful lot of time with him too – certainly a lot more than Garland did in later years after Gar graduated from high school. A neighbor boy of ours kept his young dog on his front porch and Garland saw the dog there quite

Figure 139 – Afton Bray and Dog "Tip" in 1951
Photo From Grace Bray – Tip wasn't in his preferred "element" in this photo but he did put up with handling by little kids. Our little brother Larry and also Afton both loved Tip. The real "Tip" was the eager, determined dog accompanying Gar and Eldon on a hunt. He would range thirty yards to the front or to the side, fully alert, sniffing bushes and looking for game. We were never able to break him from chasing deer or attacking porcupines but he kicked out a lot of rabbits and was a great hunting companion.

often as he walked past the neighbor's house. He also saw the neighbor abuse the dog several times. It was obvious that the boy had no affection for the dog. One day when Garland looked outside he saw the dog sitting on our porch. Nothing was said to the neighbor or by the neighbor but from that time on the dog belonged to Garland. Thus began a very enjoyable adventure of a boy (or two) and their dog roaming the hills as they hunted for rabbits and whatever other game crossed their path. Gar was about eleven years old and I was seven or eight in 1941 when he got Tip. Tip was about twelve years old when he died in 1953 – he was with both of us through the years that we grew up and hunted the hills around Copperfield.

Gar tried to train Tip to "Stay" and to "Go home" but wasn't very successful at it. This lack or only partial success at training led to some interesting incidents. When Gar started going to the seventh grade down in Copperton Tip would often follow him to the school bus down by the tunnel. Gar would tell him to "Go home!" but there were at least two times that Tip followed the school bus clear down to Frog Town and another time when he went all the way to Copperton and tracked Gar to his locker in the school. Frog Town was about 4 ½ miles and Copperton was about 7 miles from Copperfield. Luckily enough the bus driver was understanding and allowed Gar to bring Tip back home on the school bus.

Another incident – Tip had been a "bad dog" and Gar told him to stay home while Gar went hunting. This would have been very, very difficult for Tip. He would have seen Gar leave with his .22 and have known that he was going hunting which was Tip's great love. It was wintertime and the snow was about a foot deep. Gar went up the road past Telegraph and up past Big Tree (Bear Gulch) to the ridge separating Bingham from the Queen (Blackjack Gulch). Tip had apparently followed Gar's command and stayed home because Gar had walked all this way without so much as a glimpse of Tip. But then Gar turned around and started back toward Copperfield by walking northward along the ridge above his earlier tracks. He had gone about a hundred yards when, looking down at the bottom of the gulch to his left, he saw Tip following through the snow on his trail. Tip had been following him for the entire two or three miles! Gar called for Tip and they had a happy reunion and made up from their earlier difficulty.

Another memorable experience that Gar had with Tip was when they went to the butcher shop to get some bones and scrap meat for Tip. Butch would wrap the bones and dog meat up just like it was prime steak for one of his best customers. Gar had trained Tip to hold the package in his mouth and carry it home. This time, however, Gar for some reason had to go down through the tunnel to Bingham. So he sent Tip home by himself, still carrying the package of meat. Nick Golish then told Gar what had happened with Tip. Still carrying the package of bones and meat in his mouth, Tip was set upon by Pat Montoya's two dogs. Tip dropped the package onto the ground and then started fighting with both of the other dogs at the same time. He beat up on both of the other dogs, picked up the package again, and continued on home.

When Gar went into the Army in 1951 he remembered Tip in his letters. He would put a piece of paper inside his shoe and wear it for a day or two and then include it, with the notation "For Tip" in his letters to Mom and Pop. They would give the piece of paper to Tip to smell and he would wag his tail as he sniffed it.

TIP VS PORCUPINES

Tip's foremost antagonist out in the hills was the porcupine. He chased almost every wild animal that he came across – deer, rabbits, squirrels, rock chucks, but – especially -- porcupines. The reasons for his animosity toward porcupines were first, he could catch them, second, they fought back and third, they caused him pain. Every other animal that he chased was too fleet for him to catch – such as deer and rabbits. Or they would run down a hole and thus quickly escape him – such as squirrels and rock chucks.

Tip's infuriating enmity toward porcupines originated with their very first encounter. Gar and I were walking along a deer trail through the oak brush above Bishop's (the Yosemite Mine) when we came upon a porcupine that started waddling up the hill through the thick brush. Tip immediately took after it. He caught up with it in no time flat. The porcupine pushed its front end partially under the thickest part of an oak tree and turned its formidable back and tail toward its pursuer. Tip barely hesitated and then he tried to bite the porcupine. His jaws met the protective covering of thick quills along his adversary's back and, at the same time, the porcupine swatted him with its tail which was also arrayed with hundreds of quills! Tip quickly felt the pain of the quills in his mouth that resulted from his biting of the strange creature but even more painful were the quills in his nose where the porcupine had swatted him. When Tip first started after the porcupine Garland and I hollered at him to stop but he paid absolutely no attention to us. If anything he seemed to race even faster. Garland and I ran after Tip but were too late to save him from being hurt quite badly. By the time we reached our dog the battle was in full array – at least on Tip's part. The porcupine was simply holding its position but Tip was absolutely wild and had made a second and a third attempt to bite his quarry. The result had been the same as with the first bite – even more quills were now embedded in Tip's nose, mouth, and throat. Gar and I grabbed Tip and dragged him away while the porcupine stolidly maintained its position.

Poor Tip was a real mess. He must have had sixty quills in him. We took him home and used pliers to pull out the quills. You'd think that one such painful encounter would cure him of chasing porcupines but no – he became a bit more cautious but he never did stop chasing them.

TIP'S WORST ENCOUNTER

Garland and Tip were out hunting one day just below Bishop's (the Yosemite Mine). They were walking along a newly bulldozed road when they saw a porcupine cross the road about fifty yards ahead of them and go down off the road toward the hillside below. Tip, as usual, dashed after the porcupine. Gar got there just in time to see the start of the melee. Tip ran at full speed down the looser, more steeply dipping gravel of the freshly bulldozed road. The porcupine had just reached the more solid, less steep hillside. Tip tried to stop but he slid on the steep, loose gravel right into the back end of the porcupine. Sure enough, the porcupine swatted him with its tail over and over before Tip could find his footing and get away. Tip's whole chest, nose, and throat were full of quills. He was in such a bad way that he could barely move without causing such pain that he whimpered.

Garland carried Tip all the way home. He laid him down on the porch and prepared to remove the dozens of quills. Gar had heard that if you cut off the base end of a quill then the barb on the front end will deflate so that the quill can be more easily pulled out with less pain to the dog. He got a pair of scissors and some pliers from in the house and went to work. He cut off the ends of the quills and pulled them out, one by one, as Tip quivered and squirmed beneath him. Gar pulled out all the quills he could find but thought that some still remained in Tip's chest after he was finished. Gar was very worried that the remaining quills would work their way into some vital organ and perhaps even kill Tip in a few weeks. Tip lay on the porch for three days and then gradually got better and began to move around some. It was a full month before he could walk and run around without pain. The quills that remained in him never did cause him further problems but it is possible that Tip could have used his teeth and pulled them himself as they worked their way out of his body.

TIP CORNERS A ROCK CHUCK

Rock chucks can be formidable adversaries for a dog. They are considerably larger than a squirrel and they have big, sharp teeth almost like a beaver. Tip tangled with one under the worst of conditions and came out second best. We were up in the U.S. one day walking along the road by the Silver Shield mine (Porcupine Gulch). Suddenly a rock chuck ran across the road about fifty yards ahead. Tip took off after it and was rapidly catching up. The big question was whether the rock chuck would reach safety before Tip grabbed him. Fortunately for the rock chuck there was an old pipe made of wooden staves that was mostly buried just below the road. The pipe was about two feet in diameter and one of the staves was broken out. This left a hole that was just big enough for the rock chuck to fit through. This hole was probably the rock chuck's den and he had been running straight for it to begin with. He dived into it just an instant before Tip reached him.

The real battle started then. Tip was still determined to catch the rock chuck. He tried to push his head into the hole but could get it only partway in. There must have been a very limited space inside the old pipe for the rock chuck because he was within inches of the opening. When part of Tip's head protruded into the hole the rock chuck bit him -- on the end of the nose! This really infuriated Tip. With a loud growl he attacked again with renewed vigor. The rock chuck bit him again. Tip jerked his head back out of the hole with blood pouring from his nose! He yowled and attacked again but with the same results. Tip must have stuck his head into the hole four or five times and the rock chuck bit him on the nose each time. I finally grabbed Tip by the collar and pulled him away. The rock chuck was unscathed in the battle but Tip was lucky to have a nose left – he definitely was the loser.

TIP "CATCHES" A DEER

Our dog, Tip, was a black-and-white mongrel with some cocker spaniel blood for sure. Whatever other part breed he might have been wasn't obvious. But no matter – he was a great companion for a kid who roamed the hills with a .22 hunting rabbits and whatever other edible game might cross his way. Tip was especially a great help in hunting cottontails in the winter. He would go down into the brushy bottoms of the gulches and chase the rabbits while I and my partner would sit up on the side hill where we could see the rabbits if he kicked one out. Very often a cottontail would be only six feet away from him on the other side of a bush without Tip knowing it. We could plink the rabbit and then run down to pick it up. We never could teach him to retrieve or bring things to us.

But Tip had one major fault. He liked to chase deer. When we came upon a deer Tip would go absolutely wild. He would immediately charge off at top speed, howling and barking, after the deer. The deer always bounded away through the brush or up a hillside with Tip falling farther and farther behind. Tip was persistent though and sometimes he would chase the deer for half a mile or more. We would try to call him back, throw rocks at him, do everything we could think of to stop him but never with any success. It might be half an hour before Tip would come back, his tongue hanging out and a satisfied look on his face. We tried yelling "Bad dog! Bad dog!" at him when he returned but that didn't do any good either.

One day though, Tip finally met his match. I was hunting with him, walking up a bulldozer trail over in The Big Rocks. The dozer trail went just to the left of a rock outcrop that was about fifteen feet high, twenty feet wide and fifty yards long. It was plenty big enough that we couldn't see what was on the other side of it. We were about eighty yards below the outcrop when a deer walked out from behind it. Tip immediately stiffened and then, with a yowl, he took off after the deer. He was sixty yards from the outcrop when a second deer walked out. He redoubled his effort – now he had two deer to chase. Then the third deer filed out. Tip was thirty yards away and barreling along at top speed. When the fourth deer stepped out Tip was finally within a striking distance of eight feet. I wondered whether he would grab the deer by a leg, jump on his back, or leap for the throat. In all his years of chasing deer, Tip had never been in such a position as this. The golden opportunity was now his! What did he do? He put on the brakes and slid to a stop! All four deer continued bounding down the hill and left Tip standing on the ridge in their wake. He didn't even chase after them the way he normally would have!

Tip had another close encounter with a deer when he was out hunting with Gar. They were going along the ridge between Yosemite and Saints Rest Gulch when Tip started chasing a 2-point buck. The 2-point ran from Gar and from Tip for a short way but then he turned around, put his head down, and started toward Tip. This was an unexpected development for Tip and he stopped in his tracks. The deer then went on his way and Tip didn't chase him any farther.

SHELLS FOR CHRISTMAS

For Christmas every year Pop gave Garland and I each five hundred .22 shells (ten boxes). He might have had an ulterior motive because he received an awful lot of cottontails to eat for supper over the years. It was a win-win situation for all of us. During World War II meat was rationed and .22 shells were hard to get. Cottontail rabbits (and deer) helped to supplement the meat supply. We were very frugal with our shells. Five hundred .22 shells might bring over a hundred cottontails in return along with a couple pheasants and even a deer or two.

A LION IN THE BRUSH?

Lions were fairly common in the Oquirrh Mountains and during our wanderings through those hills we surely must have come very close to lions several times. I saw animals half a dozen times that could have been lions but they were so far away and ran into the brush so quickly that I couldn't be sure what they were. One instance, in which I never saw the animal at all, has especially stuck in my mind.

One day I was walking up Jackass Gulch with Tip. The brush in the bottom of the gulch was four or five feet high and very thick. Suddenly Tip growled and took off running straight ahead through the brush! I couldn't see what he was after. He barked and yowled like he was right on the heels of a cat or a deer. If he was after a deer I would almost surely have seen it bounding above the brush. I could hear Tip's barking as he ran a quarter mile up the bottom of the gulch and then it seemed he had lost the trail because he stopped barking. And then I heard another, more sinister sound. An animal was coming back down the bottom of the gulch at a fast trot. It wasn't making any sound except for its panting! I couldn't hear the breaking of any twigs or even brushing against the bushes – just the rapid panting of an animal much like Tip would do after a hard run. I followed the panting sound as it passed about fifteen yards away. I looked as hard as I could to try to see the animal through the brush but couldn't see anything at all. As it passed by me I thought "Maybe that is Tip!" so I called out "Here Tip! Here Tip!" but there was no response from the mystery animal – it just kept going and the panting sound died away down the gulch. Five minutes later Tip returned from up the gulch.

The panting that I heard was just like that of a large dog or coyote. Tip would never have chased a coyote the way he chased whatever it was. On the other hand, I have never heard a cat pant before. But I have never been close to a cat that has exerted itself either. There must be some way for cats to cool off after a hard run and panting is the only way that I can think of. So unless I find out differently about whether cats pant or not, I think it was probably a lion that ran past me that day.

THE SILENT BOBCAT

Bobcats are essentially always quiet but I had an experience with one that was especially eerie. One early Spring day I went hunting up above the U.S. (Galena Gulch). The weather was beautiful. I (with Tip) climbed higher and higher above Copperfield. The easiest walking was on the huge snowdrifts along the ridge crest. The snow had melted and refrozen several times so that it was hard and I didn't sink in. In some places a dusting of fresh snow showed a few white hare or snowshoe hare tracks and there were some pine hen tracks (either ruffed grouse or blue grouse); the animals themselves, however, were almost surely on the north side of the ridge where the snow was too deep and soft for me to walk. A few snowbirds flew by from time to time. Otherwise there were no signs of life – no deer – they were lower down the mountains where the snow had already melted. I and Tip were all alone as we walked along the ridge. We went all the way up to where the ridge formed the divide between Butterfield Canyon and the U.S.. It was past noon and I hadn't seen anything all morning so I decided to start back toward home but by a slightly different route.

I dropped down over the ridge on the Butterfield side and was pleased to find almost no snow because it was a south-facing slope. There were several vertical rock outcrops and I stopped next to one to look around. I had seen these rock outcrops on other hikes and remembered some with cavities that were partially filled with large packrat or bird nests a couple feet high. Now I saw one of these cavities in a rock face about a hundred yards away from where I stood. But there was something peculiar about it. The cliff was twenty feet high and the three-foot cavity was ten feet above the base. There was an object on the floor of the cavity but I couldn't make out what it was. It didn't quite look like a nest. Maybe it was a large cluster of cactuses? Finally I decided to take a shot at it. I rested the .22 rifle against the rock outcrop that was next to me and took careful aim. The main problem with the aim was how much elevation to give for a .22 at that distance. I decided how high to aim and carefully squeezed the trigger. All had been very, very quiet up to that time. No animals of any kind. Not even any birds. No wind. No sign of life at all except for me and Tip. With the report of the rifle the object I had shot at leaped four feet straight into the air with its legs outstretched. It was a bobcat! Then it dropped down to the ground and ran into the brush. It was unexpected and sudden and was quickly over. The hair on the back of my neck stood up. It was like a dream with no connection to reality. The bobcat hadn't made a sound and now there was nothing to be seen. There hadn't been a whisper for an hour or more until the report of the rifle and now it was profoundly silent again.

I ran over to the where the bobcat had landed on the ground and saw its fresh tracks in a patch of snow. I called Tip over and tried to get him to chase the cat down but he wasn't enthusiastic and so I gave up. There was no blood and from the way the bobcat leaped straight up it was obvious that the bullet had hit right under him. Eight inches higher and I would have nailed him. I think Tip in his early years was enthusiastic about chasing cats but maybe caught up with one and had an experience that discouraged him from chasing them in his later years.

I saw several other bobcats over the years and took a long shot at one of them but again missed. Bishop trapped one over by the Yosemite Mine. We kids never tried to trap any predators but many years later Clayton and I trapped a bobcat in Dry Fork and his friend Spot mounted it for him.

A GRUELING HIKE AND THREE CRACK SHOTS

When I was about fourteen and my brother Garland and our friend Lee were seventeen we went on a 3-day hike that I will never forget. We also did some good shooting. The three of us and our dog Tip left Copperfield with full camping gear at about noon the first day. We climbed up to the ridge separating Butterfield Canyon from the "U.S." (Galena Gulch) drainage of Bingham Canyon. Then we followed the old timber trail down to the spring near the Middle Canyon-Butterfield pass. Distance about four miles. We camped for the night by the spring and cooked something to eat before going to bed. Early the next morning we arose, had breakfast, and started up the ridge toward Butterfield Peaks. After reaching the high part of the ridge where it leveled off somewhat, we followed it to the south toward West Canyon.

That is when we had our big shootoff. As we walked out on a spur of Butterfield Peaks we scared up a covey of eight pine hens (ruffed grouse). They scattered as they took off. Most of them went downhill and we didn't want to follow them there. One bird, however, landed just below the crest of the ridge about a hundred yards away. All three of us had our .22 rifles and were determined to get that pine hen. We walked around the ridge to near where the bird had landed, laid down our packs and made sure our rifles were ready. Then, rifles in hand, we stepped over the ridge. We were each about twenty feet apart when the pine hen took off just in front of me. I raised my rifle and fired – a perfect shot! I knew I had hit him the instant I fired. The bird faltered in mid air and fell to the ground. I shouted "I got him!" But so did Garland and so did Lee – they also shouted "I got him!". Each of us, after hearing the others' claim, still insisted that his shot had been the one that brought the bird down. We picked up the dead pine hen and brought it up to where our packs were. Then we pulled the skin off and cleaned him. We all looked closely at the carcass – there were three bulletholes through him. All three of us had hit a flying pine hen with .22 rifles!

We continued around the ridge another mile or so to where it dropped off into White Pine Flat of Middle Canyon to the north and into West Canyon to the south. We left the ridge and angled down into West Canyon. It was about six o'clock in the evening when we made camp in one of the small draws of West Canyon above Fox Hollow. I was almost exhausted by this time. Garland and Lee were in a bit better shape but they also were very tired. We cooked the pine hen and fed it to Tip. We heated some canned food for ourselves. Distance covered? It felt like twenty or thirty miles but was probably about eight but with a lot of ups and downs and carrying our packs. With all the hunting and hiking I had done over the last four years I had never been this tired before. I thought that I couldn't get any more tired than I was. Little did I know what the next day would bring.

The next morning we got up early, had breakfast, and began our ascent of Lowe Peak. We mistakenly thought it was the highest point in the Oquirrh Mountains. Apparently that honor belongs to Lewiston Peak which is about seven miles farther south. We hiked southward to the main canyon that drains to the north off of Lowe Peak. We left our packs in the bottom of the canyon and climbed to the main ridge that runs directly to the peak. We ascended the ridge and made it to the peak before noon. We had done it! This had been our objective. Now all we had to do was get back home this same day.

Climbing the peak without carrying our packs had been a great relief. Going back down was even more enjoyable. There was a spring of cold, clear water that emerged from a rock face in the canyon bottom. We drank our fill and made sure all our canteens were full. Then we started for home. Not wanting to climb uphill any more than necessary, we headed for Butterfield Canyon. There were two problems, however. First, we didn't have a good topographic map and second, we only had one small canteen each. The result was that we probably took a more difficult route than was necessary, we didn't know where the nearest spring was in Rose Canyon, and we ran out of water.

We crossed West Canyon and angled up to the ridge near the top of Rose Canyon. Then we trudged up and down for hours over the small draws at the head of Rose Canyon. We must have passed close to Rattlesnake Spring without knowing of its existence. By this time we had drank all the water from our canteens, were so thirsty that we couldn't spit, and the sun was going down. We made it at last to Stocking Fork and followed it down into main Butterfield Canyon where we knew there was water. Then – after suffering from thirst for hours – we discovered that one of us had a full canteen in his pack all the while! But at least we were in what we considered to be "home territory". We had hiked from Copperfield over to Butterfield many times and knew exactly how to get home. By now, however, I was at least as exhausted as I had been the previous evening. It was already dark and we had to hike up to the Queen Mine (Blackjack Gulch), up to the ridge above Big Tree, down to Telegraph (where Lee lived) and then Garland and I had to go another 3/4 mile to the Terrace Heights where we lived.

We all three started up the road to the Queen Mine together. It was very dark so it was difficult to keep track of one another but that didn't matter anyhow. When I got so tired that I couldn't walk another step I lay down on my belly in the road and rested for ten minutes or so. Then I would get up and walk another hundred yards before laying down again. Lee was doing the same; I don't know whether he was trying to look out for me or was forced to lie down because he was too tired. Garland left us both behind and went home without us. The steepest uphill part of the road was from the Queen Mine up to the ridge. We didn't bother to take the shortcut up the pipeline (like my Dad and friends did when carrying the deer home). Lee and I struggled up the roadway intermittently and lay down several times before we made it to the ridge. It wasn't as difficult going downhill past Big Tree and the pond where the placer mine was. But it was still mighty hard. It was like pushing your body to keep going and you could just barely do so. When we finally made it to Lee's house I longingly thought of how nice it would be to just go in and go to sleep there. Lee trudged into his house and I forced myself to keep going down the road and over to Terrace Heights. It was somewhere around eleven o'clock when I made it home. I can't remember what I did then – whether I ate anything or just crawled into bed. Distance hiked that day? I'd swear it was forty miles but it was more likely twenty. Rather than distance it's probably more accurate to state the hours of hiking – about seventeen that day. After that 3-day hike I've never had to do anything that was so grueling – not in the Army or anywhere else. Some of my hunting trips were mighty tiring – but only for one day and none were as bad as the last one of our "Grueling Hike".

A WELL-HIDDEN PHEASANT

Usually I used a .22 rifle rather than a shotgun for hunting cottontails. A .22 doesn't hurt the meat as much as a shotgun, has a greater range, and the shells are far less expensive. But one day Warren and I went hunting and for some reason I took my 20-gauge shotgun. It was early in the winter and there were two or three inches of snow on the ground. We walked from Copperfield up over the saddle and down toward Lark and Horseshoe Bend (of the old High Line of the Rio Grande Western Railroad that had been abandoned many years earlier). We went a ways farther down one draw than usual and came to the edge of the wheat fields. There weren't many cottontails down that low but pheasants were more common.

Sure enough, a rooster pheasant took off about fifteen feet in front of us. I took one shot with the 20-gauge and knocked him down. But he was only wounded – one wing was broken. The pheasant started running as soon as he hit the ground and he was out of range before I could take another shot. He ran across the level ground of the wheat field and up over a little ridge a hundred yards away. Warren and I started tracking him. The two inches of fresh snow made for easy tracking. After two hundred yards we almost came upon the pheasant but he started running again before I was within shotgun range. This happened twice.

Finally, after tracking that rooster for close to a mile, his tracks went into a patch of thick sagebrush. The patch of brush was about six feet wide and ten feet long. We looked real closely into the brush but couldn't see any sign of the pheasant. We each walked all the way around that patch of brush and couldn't see the pheasant but his tracks didn't come out anywhere. He had to be in there! So we started looking harder than ever into that patch of brush. We walked round and round and jumped on top of each individual sagebrush. We looked under every piece of brush but without any luck. After more than half an hour of fruitless searching we were considering building a fire to burn him out. Then we finally saw the tip of a feather poking out and were able to shoot him again.

That was my first experience at trying to find a pheasant that was hiding in the brush. Since then I have hunted for pheasants in even smaller patches or even individual tumbleweeds or sages. It's almost unbelievable how well they can hide! Warren and I might have been lucky to have found that rooster in such a large patch of brush!

THE COTTONTAIL CIRCLED

Another odd experience that Warren and I had together was when we went hunting over by Horseshoe Bend when the snow was more than a foot deep. It was really tiring to walk through that deep snow and we had almost worn ourselves out before we started up the mountain toward home. About three-fourths of the way up toward the ridge by The Big Rocks the snow was even deeper.

Then we saw a rabbit! The cottontail was struggling through the snow much like we were and he disappeared over a little ridge about fifty yards away. I said "Let's go after him!" but Warren didn't think it was worth the chase. So he stood on the hillside, looking down the hill, while I went after the rabbit. The cottontail led me a merry chase. After going over the ridge he turned uphill and after two or three hundred yards of tracking I found myself following the tracks right back toward Warren. The tracks were fresh and clear and I got nearer and nearer to Warren but the rabbit was nowhere in sight. Finally I followed those tracks all the way right to where Warren was standing! The tracks went about five feet behind him! He had been standing still, looking down the mountain, all the while that I chased that rabbit. The rabbit must have thought Warren was a tree or a rock and paid no attention to him. He ran right past him and neither of them saw the other!

After all that walking I wasn't up to chasing the cottontail any farther. I just gave up on him. Warren and I dragged ourselves the rest of the way around the top of the mountain and home. The cottontail apparently never knew how close it had come to Warren.

GOOD SHOOTING BUT HIGH

Another strange experience happened when four of us were hunting cottontails at Horseshoe Bend. We were all walking along the fill of the abandoned railroad bed where it crossed a fairly large gulley. There was half a foot of fresh snow so that the cottontails showed up real well against it. We spotted a cottontail in the gulley right at the bottom of the fill. It was sitting broadside to us and hadn't moved at all. It was quite a ways down to the object but we were pretty sure it was a rabbit rather than a rock or tree stump; we even thought we could make out its ears. It was a steep angle down to the cottontail and we prepared to shoot it with our .22s. All four of us took two or three shots. We aimed as carefully as we could but all our shots had absolutely no effect. The "rabbit" still sat there unmoving. We finally decided it must be a rock rather than a rabbit so we continued walking toward the next gulley.

That is, all of us walked on except George Kubota. He was so sure it was a cottontail that he stayed behind and kept shooting at it. Fifteen minutes later the rest of us were hunting along the hillside in the next gulley over when George caught up to us. He was carrying a cottontail! It really had been a cottontail and it had sat stock still while we fired all those shots at him! Upon inspection we discovered that we had been shooting high and the cottontail had several holes through its ears! We hadn't realized how much of an effect that steep downhill angle would have on the trajectory of our bullets.

LOTS OF HANDS FOR DRAGGING A DEER

One year when Garland was sixteen he received a special permit to hunt deer on the Heaston Game Preserve which included the area around Copperfield and Bingham. So I went with him over to The Big Rocks which were on the top of the ridge above the Winnamuck area of Bingham on the East Side. The mine levels where waste rock was being removed had been extended most of the way up the East Side of the mine. But there was still a lot of good deer country remaining along the top of the ridge and on the hillside toward Lark. There were always a lot of deer in this area but in the fall and winter there were even more because they migrated there from the higher country and from the more northerly-facing hillsides.

Sure enough there were plenty of deer on the side of the hill where the waste dumps hadn't yet encroached. Gar was able to shoot a nice 3-point buck. He and I cleaned the deer out and then considered how to get him home. It wouldn't have been a problem to walk all the way back to Copperfield with half a dozen cottontails each but a 200-pound deer was another matter. And we would have to drag or carry the deer uphill about a hundred yards to the top of the ridge which was just above the tracks of the dump. After getting the deer to the ridge and down to the tracks it would still be a tough job to carry him the mile or so along the level of the mine to Copperfield.

Ben Valdez's #14 track gang was working on the dump and we had talked with them earlier. It was the same gang that Gar had worked on during the summer. They were all interested in our hunt and had been watching our progress whenever they could. One bunch of deer that we scared out had run right past them and the gang had heard the shots when Gar got the 3-point. Now they were curious about the shooting and whether Gar had been successful. Joe Dunn was the level foreman and he walked over to see how we were doing.

Joe immediately saw the problem we faced in getting the deer up the hill and around to Copperfield and he immediately solved it. He called the track gang over and told them to help us. Six or eight of them came down the hill and grabbed the deer by the horns and legs and dragged and carried him up the hill and to the tracks in no time flat. Gar and I were greatly relieved to have that difficult job accomplished so quickly and easily. But now we had to get the deer the rest of the way home. Joe solved that problem also. When it was time for the track shifter to leave for the day Joe had them load the deer onto the shifter. Gar and I and the deer all rode the shifter along the level and down the switchbacks to H-Level right behind our house where we unloaded the deer. We were home within forty minutes!

In those days people were more responsible for their own actions and didn't sue the companies at the drop of a hat. We were a lot freer to do things in the most logical way. The Utah Copper or Kennecott and its employees and townsfolk were far more cooperative and helpful to one another. Gar and I were responsible hunters so no harm was caused by our walking or hunting on company property. Joe Dunn and the track gang got all their work done so the little time they took to drag the deer up to the tracks didn't hurt the company but it sure helped us! The track shifter was going past H-Level behind our house anyhow so it cost them nothing to haul us along. Joe had saved Gar and I from several hours of back-breaking work for which we are still grateful.

MY FIRST DEER RIFLE

When I was fifteen years old my Dad gave me a .270 Winchester rifle. Lloyd Miller had an elk permit for the Heaston area (which included Bingham Canyon and its tributaries). He borrowed my new rifle and shot an elk in upper Freeman Gulch. The next year, when I was sixteen, I could legally hunt deer myself. My friend Warren went with me and I shot a 2-point buck in the lower part of the Left Hand Fork of Middle Canyon. A week later I went hunting with my Dad and shot a 4-point for him at the very top of the Left Hand Fork. Over the years I have used that rifle to shoot numerous deer, quite a few elk, several antelope, several black bears, one grizzly bear and a couple of mountain goats. I still have (and still use) the .22 rifle and the .270 that my Dad gave me so long ago.

ED'S FIRST DEER HORNS

Deer season was approaching. Garland's friend Ed, who he worked with, had never gone deer hunting but was anxious to do so for the first time. The two of them decided to go hunting up above the mine in Highland Boy. The uppermost mine dump on the West Side had a road along it. Immediately above the road there was real good deer country.

Ed really wanted to get a good buck on his first hunt. He planned on mounting the antlers to put on his wall and on having deer meat to eat in the coming months. He bought a new rifle with a telescope sight and all the other gear that was needed for a successful hunt.

On opening day Garland, Pop, and Ed drove up to the waste dump before daylight just as they had planned. When daylight came there was a good chance that they could spot a buck right from the road but otherwise Gar and Ed would climb up the hill while Pop kept watch from below. The sky became lighter and soon they could make out objects on the hillside. Sure enough, they saw a deer on the hill about two hundred yards above them. They looked through their binoculars and made out the antlers of a good-sized buck. The deer was standing broadside to them.

They decided that this should be Ed's buck. He was so very eager to get a deer and had gone to considerable expense and effort – he deserved to get the first deer. Ed got out of the car and sat down to get a steady aim. Then – kablam! Ed shot and the deer went down. Ed had his trophy. He immediately started up the mountain to get the deer. Gar and Pop watched from down below to make sure the deer didn't get back up. Ed climbed as rapidly as he could and soon he could see the deer. It was still lying down but wasn't dead. Ed approached the deer from the front and decided to finish it off by shooting it in the head. He aimed carefully and shot. Kablam! He had killed his first deer. Much to his astonishment and disappointment, however, the horns fell one to each side and dangled there loosely.

The trophy set of antlers Ed had anticipated proudly displaying above his fireplace now looked more like something from a comic strip. He couldn't even use the horns to drag it down the hill. Gar climbed up through the brush and helped Ed clean out the deer. Then they each got hold of a hind leg and pulled the deer down to the car. In the end Ed, with some ingenious taxidermy work, was able to have the two antlers rejoined and mounted as a trophy.

BEARS AND LIONS IN BINGHAM

The entire Oquirrh mountain range, including Bingham Canyon, has an abundance of chokecherries, acorns, service berries and other food that bears just love to eat. There surely must have been a goodly population of these animals in the early years of Bingham. Lions still live in the Oquirrhs and they roam the parts of Bingham Canyon that are away from the active mines and dumps.

Following is a story that was provided by Eugene H. Halverson.

Edla was a relative of mine who came to Bingham as a young lady in 1903 and returned to Finland in 1913. She wrote the story in Finland. The book was written in the Swedish language and published by K-G Olin in Finland. "No fa ja te America!"translated from Klippiga Bergen by K-G Olin.

Now I am going to America!

The boardinghouse was located in the canyon between the mountains and to reach the city itself, you had to walk through the mountain. On the road to and from the dances the girls were afraid to be attacked by mountain lions. If you run into one you were supposed to stare them straight in the eyes, and they would leave you alone. It would stay still or run away. Edla was told that mountain lions had attacked some from behind and killed them.

During the nights they always kept a gun or an ax close to the bed because there were bears and rattlesnakes. One night when Jansson went over to another claim to borrow some light, he met a bear. Jansson rushed down the mountain to Bingham. Only the next morning did he dare come back up the mountain.

The Canyon was also something special. Another Osterbotten immigrant described the area as "one of the most unusual the Lord and people had ever put together". What made Bingham so unusual was that it was squeezed in along a long winding road that went along the side of a sharply sloping canyon wall. That made the houses look like they were stacked one on top of another up along the mountainside. The community was described as 10 Kilometer long draining ditch. Another reason that made Bingham special that out of the 20,000 population there was not less than 30 nationalities represented. A large group was made up of Swede-Finns, at the time of Edla's arrival and was between five hundred and a thousand. Most of the men from Osterbotten worked in the mines.

ENTIRE CAMP UNDER STRICT QUARANTINE

At Meeting of Town Board Entire Camp Is Placed Under Quarantine – County Physician Approves Action – All Persons Leaving Camp Must Secure Permits In Order to be Able to Return – Meeting Attended by Large Number of People.

Bingham has quarantined against Salt Lake and other outside districts on account of the epidemic of influenza and the further fact that the city of Salt Lake has lifted the ban and permitted the opening up of picture shows and other places where crowds assemble, notwithstanding the fact that Salt Lake is literally permeated with this dangerous disease.

The ban in Salt Lake was listed last Saturday and on the afternoon of the same day the town board and the citizens of Bingham came together and decided that the continuance of traffic between this camp and that city would be a serious menace to the health of the people here, and to guard against such dangers the entire camp including Copperfield and Highland Boy was quarantined, and the regulation became effective at noon last Monday.

After getting the views of the people assembled and the approval of the county physician the following proclamation was issued by Mayor Q. B. Kelly;

"Whereas, the regulations requiring the wearing of masks and the efficient labors of our physicians and nurses have reduced influenza cases to the minimum, and,

"Whereas, the epidemic is raging on the outside and fatal cases have been traced to contact of our residents with people living elsewhere, and

"Whereas, at a meeting of the Town Board attended also by a large number of citizens a resolution was passed and unanimously adopted quarantining Bingham and

"Whereas, Dr. Howell, the county physician, was called in consultation and approved the action of the said meeting insofar as the county districts of Highland Boy and Copperfield are affected.

"Now, therefore, I, Q. B. Kelly, mayor of the town of Bingham, by authority vested in me by the constitution and the laws of the State of Utah, hereby proclaim and declare beginning at 12 o'clock noon, on Monday, December 9, 1918, all residents of Bingham are prohibited from leaving the district unless they choose to remain away until this ban is lifted, and all outside people are prohibited from coming into Bingham without first having secured a permit from the town health officer, or from a resident deputy sheriff when the case applies to Copperfield or Highland Boy, excepting that the traffic of physicians and nurses shall not be interfered with in the practice of their profession.

"Given under my hand and the Seal of the Town of Bingham this seventh day of December, A. D., 1918."

With this new regulation there has been little traffic between here and Salt Lake since Monday at noon. The influenza situation here has been steadily improving, and it is hoped that the disease will be entirely stamped out by the methods adopted.

CHAPTER TWELVE

AFTER HIGH SCHOOL

After graduating from Bingham High School I continued working on the track gang for Kennecott. Two other young men on the gang encouraged and helped me to get registered to go to the University of Utah that Fall. My major was Geology but only a few of my classes were in that subject. By working summers and weekends and holidays while going to school I was able to pay all my expenses and have a little to spare. I went to school for two years before going into the Army for two years. Upon returning from the Army I talked to the Chief Geologist, Russell Anderson, who gave me a job as a drill sampler. I owe thanks to Russell for giving me that job. Working as a churn drill sampler led to an odd happening that I have never forgotten.

HE JUST STUFFED A RAG DOWN THE HOLE

Kennecott was drilling a churn drill hole in a very strange location – on the concrete foundation of the old abandoned machine shop by A-Bridge in Carr Fork. Ray Watson and Dee Christensen were the drillers, Johnny Osoro was the helper; I was the sampler. The drill rig was set up on the concrete floor of the shop. All the upper walls and the roof had been torn down but the two-story basement still remained. The big open room in the basement was being used as the sludge pit. The stairs to the bottom of the basement went past a room halfway down that was set up for the sampler to dry and bag the samples. The other small room at the bottom of the stairs was blocked off from the sludge but was vacant (I thought).

A churn drill makes a hole by a cable lifting and then dropping a very heavy bit down onto the ground that it is penetrating. The bit may be two feet in diameter and five feet long and is made of iron or steel that has a four-lobed cross-section. A pipe, maybe 30 inches in diameter, is set up vertically over the point where the hole is to be started. The bit is then lowered by a cable into the pipe. The drill rig then lifts the bit a couple of feet and drops it several times per minute. As the bit hits the bedrock it smashes it into small pieces or even to a powder-like consistency and the hole is made deeper. The hole also has water in it. After making several feet of new hole the bit is pulled out and a bailer is lowered to the bottom and then pulled up with a load of mud and broken rock in it. The bailer was discharged, for this drill hole, into a chute that went through a splitter and then on down to the basement floor.

My job was to collect a representative sample of the sludge from the splitter, take the sample into the sampler's room, dry it in a tub over a heating element, and then bag it. This work with the

Figure 140 – Churn Drill Rig On Old Shop Foundation About 1953
From The Gordon Bodily Collection - The drill rig is in the center of the photo. It is sitting on the upper floor of the old shop building which housed the Machine Shop, the Blacksmith Shop and the Boiler Shop. The basement of the old shop served as the sludge pit for the drill cuttings, or mud. A-Bridge and Carr Fork are to the right. The open-pit mine is to the left of and behind the drill rig.

sludge and dried samples created a lot of dust. I would sweep up the dust and dispose of it in the easiest and simplest way. There was a one-inch hole in the middle of the floor – it let out into the empty, unused room immediately below. I would sweep up all the dust into a pile over the hole and then sweep it back and forth until all the dust had disappeared down the hole, never to be seen by anyone again.

One day, after the rig had been running for a month or so, I was busily sweeping the floor and disposing of the dust by sweeping it down the hole. Out of the corner of my eye I saw movement. Looking up, I saw an old electrician slowly climbing up the stairs from below with a determined look on his face. He reached the level of my room and came toward me but seemingly without even seeing me. I stood still and waited for him to speak. But he didn't. He didn't acknowledge my presence in any way. He just pulled a rag out of his pocket, bent over, and stuffed the rag down into the hole, tamping it in quite solidly. Still without a word or even a glance at me he turned and walked back down the stairs. He had accomplished his task and delivered a message in very plain language. I understood that the room beneath mine was not vacant and that I should stop sweeping the dust down the hole!

EATING TWELVE FISH PER DAY

One summer, after I was married, Garland, Dick Steinfeldt and I backpacked into the Uintas for a fishing trip. Gar and I had previously fished Spider Lake and found it to be good so we decided to go there. Gar could only go for three days while Dick and I could stay for five. Each of us packed our own gear and food for ourselves. We left our homes in West Jordan early in the morning, drove to Duchesne, turned north to the Uintas and went up Yellowstone Creek to the end of the road. We parked the car, put on our packs, and started hiking up the trail about 10:00 a.m..

On a previous trip to Spider Lake my friend George and I had made the mistake of going up the Swasey Hole trail which also led to Spider Lake. The problem was that the trail was extremely steep and reached the 10,500-foot elevation very quickly and we weren't acclimated to the high altitude. Then the trail went up over a ridge and back down and then up over another ridge and back down so many times that we wore ourselves out by puffing up and down at that high altitude. We had to rest quite often and hiked more slowly than we planned on. Then it started to turn dark. We didn't know how much farther it was to the lake and we were extremely tired so we decided to camp where we were for the night. This section of the trail led steeply down a rocky ridge and the only level spot was right in the middle of the trail. Almost in desperation, we laid our sleeping bags out in the middle of the trail and went to sleep. Luckily for us nobody came riding horses along the trail in the dark. The next morning we got up and were packing our gear when I saw George throw something behind a bush. I went over and looked behind the bush and was amazed to see a 2-ounce package of Brillo pads! We both had been mighty tired the night before but this was a real act of desperation. George was trying to lighten his load and all he could find to dispose of weighed just two ounces! We resumed our hike and it took less than an hour to reach Spider Lake. We camped there for two days and had a good time fishing but swore we would never again take the Swasey Hole trail to Spider Lake.

Garland, Dick and I followed the trail up Yellowstone Creek. It wasn't steep at all and was at a lower elevation than the Swasey Hole trail. It was easy going. We were sure that we could make it to Spider Lake well before dark. We hiked up the trail for about twelve miles until we came to the small creek that ran out of Spider Lake which was a mile and a half up the mountain to the west. We stayed by the creek as we climbed up the mountain. We saw numerous little brook trout in every hole!

We finally couldn't restrain ourselves so we laid down our packs and started fishing. The fish were only about seven inches long but there were literally hundreds of them! And they were so ravenous that they would fight for the bait and we caught a fish on almost every cast! In no time flat we caught about sixty of the little brook trout and figured that was enough for us to eat during the next four days.

We hiked the rest of the way up to Spider Lake and set up camp on the shore of the lake. It was well before dark and the hike in had been much easier than it would have been on the Swasey Hole trail. Each of us cooked our own supper. We each had a couple of the fishes and Dick and I had cans of beef stew. Garland had opted for dehydrated food so he mixed up some soup to eat with his fish.

The three of us camped and fished for another day and a half and had a good time. We saved the fishes that we caught out of the lake because they were considerably larger and we ate some of the tiny ones from the creek. Dick and I continued to eat the canned food that we had brought with us and Gar continued to eat his dehydrated food. After lunch on the third day Gar had to leave for home. He still had a lot of dehydrated soup and Dick and I were running low on canned goods so he gave us all of his remaining soup. Then Gar left for home while Dick and I stayed at Spider Lake.

We had the last of our canned goods for supper that night. In the morning Dick and I fried a couple of fish for each of us and made up a pot of dehydrated soup. We ate the fish and started to eat the soup. Then we decided that a couple more fish apiece would taste good so we cooked and ate them. That was four fish each for breakfast. At lunchtime we went through the same routine – we had a couple of fish and a small taste of dehydrated soup and then cooked another fish each. That made seven fish for the day. At suppertime we didn't fool around with the soup. We just cooked five fish for each of us and that was the main course. Total for the day – twelve fish each! Those little brook trout taste better than other trout – and they especially taste good when compared to dehydrated soup!

ADVENTURES AS A GEOLOGIST AT BINGHAM

After graduating from the University of Utah I went to work for Kennecott Copper as an Industrial Engineer. After six months there was an opening for a Geologist so I transferred to that position. The work that I did was primarily geologic mapping of the pit, logging of diamond drill core, and microscopic examination of igneous rocks. This work was all very interesting to me.

SHAKING - FROM FRIGHT OR COLD?

While working for Kennecott at Bingham I had one adventure the results of which could have been disastrous. One of the goals of the Geology Department was to gain a more complete understanding of the intrusive igneous rocks of the mine and surrounding area. The open-pit mine was largely surrounded by underground workings and several of these intersected igneous rocks that were of interest to us. It was necessary to go into many of these old workings to collect samples of these rocks and to map their thickness, the direction they were striking, etc. These samples were taken back to the office and examined under the microscope and used to construct a comprehensive map of the Bingham District.

The Utah Metals Tunnel had one entrance in Highland Boy and the other in Middle Canyon. It had several bodies of igneous rocks that I wanted to map and get samples from. I sometimes went into the Utah Metals Tunnel by myself but most often two of us went together for safety purposes. Oxidation of sulfide minerals and the rotting of timbers consumed oxygen from the air of the old workings. If the ventilation was extremely poor the air could become so depleted of oxygen that it wouldn't sustain life for a man. The location and amount of bad air in an underground working could change from day to day as the weather (and hence air pressure) changed outside. A mine could "breathe" bad air in and out so that on a day of high air pressure the bad air would be inhaled back to the very deepest recesses of the crosscuts. However, on a day of low air pressure the bad air would be exhaled out to fill the crosscuts almost all the way to the main tunnel.

There was some bad air in the Utah Metals so we used lighted candles to gauge whether the air was breathable or not. The main tunnel itself had no bad air. Two of the crosscuts, however, were especially questionable. In one crosscut the air was good near the main tunnel but would become progressively "stale" the farther we went so that the candle would burn lower and lower until the flame was only half an inch high at which point we would retreat. In another crosscut there was a small dam for gathering the water. The dam was about four feet high and the water behind it was two feet deep. The candle would burn fairly well as we approached the dam but the flame would immediately go completely out when the candle was lowered behind the dam.

There was a vertical shaft in the Utah Metals Tunnel that extended from the level of the main tunnel to a depth of a hundred and fifty feet or so. Cold water dripped from the walls of this shaft and formed a deep pool at its bottom. It was used as a source of drinking water by the town of Bingham.

Old geologic maps showed that the shaft bottomed in igneous rock. On the day of this "adventure" my partner, Paul, and I had determined to get a sample of the rock from the bottom of this shaft. We parked our Jeep near the Highland Boy entrance and walked half a mile in to the shaft station. The shaft itself was twenty feet to the side of the tunnel. At the top of the shaft there was a platform of timbers and heavy boards and a pump and an electrical substation. A twelve-inch water pipe and a vertical ladder extended from the top all the way down to the bottom of the shaft. Paul and I looked the situation over and decided that one of us (Paul) should stay at the top of the shaft while the other (me) descended the ladder to collect the rock sample.

We hadn't bothered to talk to anybody in the Water Service to determine the condition of the shaft or of the ladder, how deep the pool of water was at the bottom, whether there was a place to stand at the bottom, or whether the air was good down there. We figured that if the Water Service people could go up and down the ladder then we could too.

I was equipped with a hard hat, electric cap lamp (with belt and battery around my waist), an army ammo bag over my shoulder, and a rock pick thrust through a holder on my belt. No rainproof slicker or warm clothes – just my regular summertime dress of shirt, pants, and leather boots. And no candle with which to check the air (I wouldn't have had a hand free to hold it and the water falling down the shaft would have doused it anyhow).

I started down the ladder and quickly began to wonder if this was a bad idea. The ladder was absolutely vertical and climbing down it was an entirely different proposition from climbing down an inclined ladder. I couldn't just stop and rest my weight on the ladder without hanging on tightly with my hands or arms because I would fall off backwards if I did. Besides that, there was no way to see the ladder below me. The light was on my hard hat shining in whatever direction my head was pointing and I was clinging so tightly to the ladder that there wasn't room to tip my head down. Near the top of the shaft there was only a light film of water on the rock walls but the amount of water increased with each foot that I moved downward so that by the twenty-foot mark the water was falling as in a light drizzle. The falling water made the ladder wet and slippery so that it was necessary to very carefully make certain both feet were firmly placed on a ladder rung and both hands were tightly gripping the ladder before moving down another step. The quantity of water falling into the shaft steadily increased with depth so that by the fifty-foot mark water was coming down as in a good rainstorm. And then, as I reached downward with my foot to get a purchase on the next lower rung, there wasn't one! I anxiously groped up and down and to the sides with my foot until, by extending myself considerably, my foot contacted the next rung down from the missing one. At this point (and several others as well) I considered going back up the ladder and giving up the whole enterprise. But I kept going down and the amount of water that was pelting down the shaft kept increasing and I was thoroughly soaked and getting cold. There were a couple more missing rungs in the ladder but I continued descending and finally reached the bottom of the shaft at which point there was an absolute monsoon of frigid water cascading down. Luckily enough, at the bottom of the ladder there was a pile of small rocks that projected above the pool of clear water into which the pipe extended. I stepped off the ladder onto the pile of rocks. I shined my headlamp around and saw that there wasn't any kind of a station down there – the entire space was no bigger than the size of the shaft. And I was still breathing okay so the air must be alright. I turned to the rock wall, looked to make sure it was igneous rock, and hammered out a couple of fist-sized samples with my rock pick. After putting the samples into my shoulder bag and replacing the pick, I contemplated what to do next.

There was no shelter from the constantly falling water – no way to warm up. The longer I stayed down there the colder I would get. The only thing to do was to climb back up the ladder as quickly as I could. The condition of the ladder was by now well known – it was dangerous but would support me and could be climbed if I was careful. So I started back up. The missing rungs in the ladder were obviously evident as I climbed up (they apparently had been broken out by falling rocks) so they were negotiated without as much anxiety as I'd had in coming down. Even so, the wet, slippery, vertical ladder and the cold water made for a perilous ascent. I clung to that ladder with my hands, feet, and even my arms, elbows, and knees. It was almost as if my body was glued to it. I cautiously moved only one arm or leg at a time and only after making certain that my other arms and legs were secure. I was impatient to reach the top but I had to be extremely careful! It seemed to take forever to climb twenty feet. After an interminable time, however, the amount of falling water eased off to almost nothing and the hum of the pump was evident. It was with enormous relief that I clambered up the last step of the ladder and onto the thick, solid planks of the platform.

It was then that I started shaking uncontrollably. It was a good thing that I hadn't started shaking until I climbed out of the shaft. If I had started shaking while still on the ladder it would have made the climb even more hazardous. I had been both cold and frightened for the entire previous hour and to this day I don't know whether the cold or the fright was the greatest cause of my shaking.

Paul was waiting at the shaft station for me and we wasted no time in heading out of the tunnel. The bright sunshine and warm air that greeted us as we emerged from the tunnel were very welcome. I felt as if I had narrowly escaped a terrible doom. It was obvious that the Water Service workers hadn't gone down that ladder in a long, long while. And when they had gone down they might have roped themselves off or used safety belts. I never did ask them how they did it or how many years it had been since anybody had climbed down that ladder but I expect it had been quite a few.

DRAGGED IN A MINE BY A "BICYCLE"

While working in Kennecott's Geology Department our accident-prone fellow employee had a very ridiculous misadventure. As a break from our day-to-day routine and to widen our horizons in the geologic field, our boss, Fred, had arranged for us to visit an underground mine in Park City. We four Kennecott visitors met with two of the staff of the United Park City Mines early in the morning. Each of us was provided a hard hat with an electric lamp and belt with an attached battery. Then we were introduced to our means of locomotion into the underground workings – "bicycles" with four heavy iron wheels with which to ride along the small train tracks.

The main adit or tunnel was more than a mile long and served to drain the water from the mine workings. The tunnel was considerably larger than most underground workings but was still a bit congested. A large stream of water ran along the bottom of the tunnel. A narrow gauge railroad was emplaced on wooden planks above the stream. Small ore cars and our "bicycles" rode upon the rails with we visitors sitting atop the "bicycles". The "back" or roof of the mine was largely unsupported – there were no wooden beams – just jagged rocks that had been unevenly trimmed so that some projected downward into the tunnel more than others. There wasn't enough vertical space everywhere for us to sit upright on the bicycles. It would be necessary to hunch down at times lest we crack our heads on a rock projecting downward from the back.

The threat of cracking our heads was sufficient to arouse our "safety" man to issue dire warnings. He had worked for more than a decade in underground mines and was familiar with all the hazards that might be encountered. This was an unusual, very obvious menace that we had to guard against. We had to crouch down on the "bicycles" and carefully look for protruding rocks before raising our heads as much as an inch from their lowered positions.

And then we were off! It took a lot of hard pedaling to get the "bicycles" moving because the iron wheels were so heavy. On the other hand, once they were moving it was hard to stop them. All six of us got our "bicycles" up to a speed of ten or fifteen miles an hour after which we could maintain the speed quite easily. We could relax a bit and look at the scenery of the rock in the walls and the back. Of course all of us kept an especially sharp eye on the back which was zipping by just a few inches above our heads.

All was going well. I was just ahead of our well-experienced "safety" man and was enjoying the ride when a loud "Whap!" sounded behind me. Looking back, I saw our "safety" man, bent over backwards, being dragged along the track by his "bicycle". His feet were caught up in the pedals and his safety hat and lantern, being attached to the belt and battery were also being dragged loosely along with the light shining uselessly in all directions as it bounced along the track. He, of all people, had raised his head up and been smacked at the top of his forehead by a rock projecting down from the back! All of us had been alerted by the loud report and watched as the "bicycle", with the momentum of its heavy wheels, dragged him for twenty or thirty yards before it stopped. Those behind him slammed on their brakes and managed to stop without running over him. There was some fear that he might untangle his feet from the pedals and drop into the swiftly moving water below but his feet remained securely held by the pedals.

Our "safety" man was, luckily, uninjured. The hard hat had protected his head from the enormous wallop he had received. He also escaped injury from being dragged along by the "bicycle". The only real injury was to his pride so little was said of the mishap but it still brings belly-laughs whenever it is recalled.

THE HYDRANT IN THE STREET

Practically all the streets of Bingham were extremely narrow because the canyon bottom itself was narrow with the hillsides rising steeply on either side. There just wasn't enough room for a two-way road with minimal parking space plus buildings on both sides. The main business section of Bingham, just above and below Carr Fork was one of the worst stretches. Traffic jams were common with the uphill traffic having to wait while the downhill traffic cleared an obstructed stretch of road. The honking of horns and sometimes violent arguments were part of these traffic jams. "You back up! No – I was here first - you back up!" was often heard. Cars trying to inch past each other often incurred scratched paint jobs and scraped fenders. Greyline buses carrying tourists to see the mine, being so much larger than cars, caused especially difficult traffic jams.

The canyon was so congested that there wasn't room for all the fire hydrants except to put at least one out in the street next to the sidewalk. One hydrant in particular, a hundred yards below Carr Fork, was two feet out in the street from the sidewalk. It was a real menace. There was a concrete curbing about fifteen inches high that was built around the hydrant to protect it from automobiles.

As the open-pit mine expanded it inexorably chewed away at the town of Bingham. This continued until Kennecott purchased essentially the whole town. The populace moved down to the valley and the buildings were demolished one by one. The roadway remained for quite a few more years, however, being used by the workers at the mine. There were no traffic jams – just an open road with very little traffic except at shift changes. But the menacing fire hydrant with its protective curbing still crouched in wait.

One bright summer day three employees of Kennecott's Geology Department were going down the canyon in a pickup truck. The pickup had a half-inch sheet of iron plate covering the bed. The iron plate was for added weight so that the truck would have more traction and wouldn't get stuck in the winter snow. The cab was a bit crowded so one man rode in the back. This individual loved to soak up the sunshine so he took off his shirt and lay down on his back on the iron plate of the truck bed with his arms outstretched. The driver was an extremely safety-conscious man who, nevertheless, was very accident-prone. There was no other traffic at all and they confidently zoomed down the deserted roadway at about twenty-five miles per hour.

Suddenly – kabam!! The front end of the pickup was thrown violently up into the air and to the left and then – kabam!! The rear end of the pickup was likewise thrown up and to the left. The passenger who was lying on his back on the steel plate and was half asleep received a tremendous double jolt. His prone body was flung half a foot into the air by the first impact and then, just as his body fell down to the iron plate it was whammed by the second impact. What a rude awakening!! The driver exclaimed "What was that!!??" He hadn't even seen the fire hydrant and its protective curbing. The only one who knew what had happened was the front seat passenger who had helplessly watched as the pickup was driven right at the fire hydrant.

The curbing did its job in superb fashion. Even a head-on collision by a pickup truck had left the fire hydrant untouched. The pickup sustained one flat tire and a bent wheel rim. The passenger in the rear of the truck had a sore back and a bruise on the back of his head that hurt for several days. The major damage was to the driver's ego. He tried to keep the incident quiet and managed to do so for several days before it leaked out. The narrow streets of Bingham Canyon had claimed one last victim.

THE NEVER-EMPTY WATER CAN

The Gemmell Club was the center of entertainment in Bingham for many years. It was a large building that had a gymnasium, stage, and lounge on the main floor and a boxing ring, pool hall, bowling alley and showers in the basement. The bowling alley was in the center of the basement. There were two bathrooms in the basement – one on each end of the building. The Club also had an upstairs floor with a library at one end; rows of seats for spectators occupied one side of the area above the gymnasium. Generations of Binghamites spent many hours in the Gemmell Club.

After most of the town of Bingham had been demolished the company converted the Gemmell Club to its uses. The large lounge area on the main floor was made into offices for the Industrial Engineering Department. The library became offices for the Geology Department and the bowling alley became the drill core storage area. The east side of the bowling alley was sealed off completely and a padlock was placed on the door on the west side. A large door was built into the south wall with a ramp that led from the roadway to the door and down into the bowling alley. Rows of racks for holding drill core, a sturdy bench for examining the core and fluorescent lights were installed above the bench and in the ceiling.

The procedures for handling the drill core were that a truck driver and helper would go to the drill rigs on the weekend, load all the boxes of new drill core into the truck and haul it to the old Gemmell Club. They had a key to the outside door to the core storage area but not to the rest of the building which was locked because nobody in the building worked weekends. The truck driver and his helper would use a cart to move the core from the truck into the building and would stack the boxes of core next to the bench after which they would turn off the lights, lock the door and leave.

Figure 141 – Gemmell Club With Ramp to Core Storage About 1960
From the Gordon Bodily Collection - The entire Gemmell Club was taken over by Kennecott in about 1957 for its own use. The Geology Department used the old library that was upstairs for its offices and the bowling alley in the basement was used for drill core storage. In this photo the ramp

to the core storage area is on the extreme left bottom of the photo. The ramp let into the large room where the bowling alley used to be. This was the site of the "Never-Empty Water Can".

We geologists would come to work on Monday and the core would be waiting for one of us to log. The procedure was to turn on the lights, lay the core boxes on the bench in sequence and then to log the core. To log the core the first thing the geologist did was to break off a piece about six inches long. Then, using his left hand, he would immerse the piece of core into a coffee can full of water and rub the drilling mud off it. The wetness also made the individual mineral grains stand out better than on dry core. Then the geologist, with a hand lens in his right hand and piece of core in his left, would hold the core up close to the light and peer at it through the hand lens. The core almost touched his nose during this close examination. He would write the footage and description of the core onto a form and then continue with the next section of core until he had logged all the core that had been drilled the previous week. Then he would dump out the remaining water from the coffee can, place the can on a shelf, turn out the lights and lock the door leading to the rest of the building.

Ed, the geologist who had newly taken over the drill logging assignment, had asthma. He periodically would pull out an inhaler and squirt some medication up into his nose to relieve the effects of the asthma. One day Herm, the Division Geologist, went down to the core storage area to look at a particularly interesting interval of drill core. Ed was already there logging the core so Herm joined him. As Herm held a piece of freshly-washed core up to his face his nose was assailed by a particularly pungent odor. He made a sharp exclamation and asked Ed what was in the coffee can. Ed told him it was water. Herm looked into the can and smelled the liquid. And it sure wasn't water!

It turned out that Ed had been logging the core for weeks and then, if he thought of it, emptying the can each time and placing it on the shelf. The next week he would prepare to log the core again and note that he had forgotten to empty the can the week before or it had been magically refilled because it had considerable "water" still in it. He would proceed to rub the mud off the core in the can, hold it up to his face, and examine it. There was no problem. With Herm's discovery, though, the real truth came out. The truck driver and helper would bring the drill core into the storage area. Then, feeling the call of nature, they would find the door that allowed access to the restrooms to be locked. But there was an empty coffee can so they made use of it. Thanks to Herm's ability to sniff things out the mystery was solved of how Ed's coffee can was never empty.

IT'S BETTER TO BE TALL IN A RIVER OF MUD

It was quite often necessary to go into some of the old underground workings to map the geology, collect rock samples or, in this case, to measure water flows. Rick wanted to determine how much water was running out of a crosscut in the old Armstrong mine. One had to be very careful in these old workings because of the hazards of falling rock, rotting timbers, bad air, etc. So it was prudent to have a second man along just in case. Rick chose to take Larry into the Armstrong with him. They made quite a pair. Rick was tall with long legs. Larry was short with short legs.

Rick and Larry gathered all the gear they might need – hard hats and cap lamps, rock pick, sample bags, note book and pencils, a map of the Armstrong tunnel, etc and off they went. They wore knee-high rubber boots because they anticipated having to wade through a few inches or even a foot of water. Little did they know what awaited them.

They drove in the Jeep to the tunnel entrance, parked, and were soon making their way through the old tunnel. Nobody had been in parts of the workings for years or even decades. Oxidizing sulfide minerals had formed brown coatings on almost everything, slowly dripping water fell from the back and ran down the walls, and there was an inch or so of slimy mud on the track. But these conditions were about what they had expected and they slogged along in their rubber boots without any problem. They made their way through the old workings for almost an hour. They followed the map along the main tunnel and then along drifts or crosscuts until they finally reached their destination which was a crosscut where they intended to measure the water flow.

A problem was immediately obvious. There was a cinderblock wall blocking the way! The entire opening to the crosscut had been sealed off years ago with cinderblocks – apparently for the purpose of controlling the air circulation in the old mine. Rick was determined that he would measure how much water would flow out of that crosscut. There would probably be a flow about a foot across and a couple inches deep. The only thing to do was to knock a hole through the wall. He and Larry went to work with their rock picks, hammering and picking away at the cinderblocks at the base of the wall. Finally they loosened a couple of the cinder blocks and Rick was anticipating success. He wiggled the bottom three cinderblocks and was able to wedge his rock pick between two of them. Then, using the rock pick as a lever, he was able to force one of the blocks out one inch and then another until he was on the verge of removing it entirely.

With one final effort Rick levered the block all the way out. The response to his success was an immediate pulse of thick, viscous, orange-brown mud from the opening. Then, as he and Larry watched in alarm, a couple more cinderblocks were forced out by the now-gushing mud and the flow increased to startling proportions! They both leaped back but the onrushing mud quickly overtook them. The mud in the 5-foot wide crosscut quickly rose to halfway up their boots and was still rising! They both backed against the side of the crosscut where there was a small pile of rocks that had built up over the years of spalling from the sides. But the mud kept coming – increasing in volume and depth as it flowed past them.

There was only one thing for Rick to do and he was barely able to accomplish this! He leaned out, reached upward with his arms and stretched his body the full width of the crosscut so that he was firmly braced in place with his feet against the lower part of one side while his hands were higher up on the other side of the crosscut. By holding his feet on the very highest inch of the loose rock pile he was barely able to keep the top of his boots above the rushing mud.

Larry, however, was not so lucky. He was too short to brace himself between the two walls and the pile of loose rock wasn't wide enough or firm enough for him to stand on. He frantically tried to think of something to do to save himself but there was no escape. The slimy brown mud steadily but quickly rose up and then over his boots and in the end there wasn't anything he could do but stand with his feet widespread to hold his balance. The mud filled his boots and continued to rise until it was a couple inches above his knees.

Rick held himself in place while Larry fought to maintain his balance as the river of mud flowed by. It would have been horrible if either of them had fallen face-down into the gooey mess. After a few minutes the flow of the mud began to abate. The depth of the mud could be easily measured by how high it was on Larry. When the mud had dropped until it was four inches below the top of Larry's boots then Rick figured it was safe for him to come down from his precarious position.

The two of them gathered themselves and waded through the mud back to the main tunnel. The mud was by then only a couple of inches deep but there were marks of the fresh mud for a foot up the sides all the way to the portal. The flow of mud had preceded them all the way and was now running down the side of the road. They had to be very careful getting into the Jeep without smearing mud all over the seats but they made it back to the office where they hosed the mud off their boots (and Larry off of his pants and legs).

It was an unexpected adventure for them but no real harm had been done. They laughed about it but wouldn't want to repeat it again. The brown mud had been ponding up in the crosscut for years and years until Rick and Larry had finally released it all at once. They learned one lesson in particular – it's better to be tall in a river of mud.

SCOUTMASTER ADVENTURES

Several years after I was married and working for Kennecott I moved my family to Copperton and was appointed to be the Scoutmaster. Bud Patrick was the very able Assistant Scoutmaster. While working with the scouts I witnessed several very humorous incidents involving the young boys.

RETRIEVING MIKE'S TACKLE BOX

With Bud Patrick

The first Spring fishing trip of the Copperton Boy Scout Troop 150 to Kidney Lake in the Uinta Mountains was scheduled for the last week in June. This was quite early in the season and we would probably be the first fishermen to go into that area. For example, the highway to Mirror Lake (which was at a slightly higher elevation) normally had to be plowed open for the Fourth of July weekend. Getting caught in a snowstorm or finding that all the snow hadn't yet melted at Kidney Lake was a possibility. However, the winter hadn't been especially severe so we went ahead with our trip as scheduled. On Monday afternoon we hiked into Kidney Lake and began to set up camp near the creek that ran out of the lake. The site was free of snow and totally dry. The wrangler packed most of our gear in on horses and we soon had our tents pitched and had a comfortable camp.

Bud and I walked up to where the creek came out of the lake and saw that there was a wooden structure with a wheel and a long iron rod extending way down in the water to the valve that controlled the flow going out through a culvert into the creek. We looked the situation over with some consternation because it appeared to be quite dangerous. The level area or platform on the top of the structure was just big enough for one or maybe two people to stand. There was no handrail or anything to hold onto. We had to be extremely careful to maintain our balance and not fall off. The surface of the lake was four feet below the little platform. There was a measuring stick that showed the depth of the water to be nineteen feet. The scary part of it was the thought that if anyone fell into the lake he might get sucked part way into the culvert and not be able to get out. And the water was so cold that a person wouldn't last very long in it. We looked out into the middle of the lake and saw a big "island" of slushy ice that was close to a hundred feet across. The water had to be close to freezing.

We decided that the structure at the outlet was too dangerous for the boys to fish from. We told the boys that nobody was to go onto the outlet structure. They all heard us. They all agreed. So what did they do? The very next day Mike went out onto the platform at the outlet to fish.

Mike had brought his Dad's tackle box with him. There was a real variety of lures, hooks, sinkers, etc in it. They must have been worth over a hundred dollars. So guess what – he accidentally knocked the tackle box off the little platform at the outlet and it sank clear to the bottom in the cold, cold water just a few feet from the culvert through which the strong current of water flowed. The kid was really distraught and was on the verge of tears. His Dad had entrusted his tackle box to him and now he had lost it! When Bud and I heard what had happened we tried to calm him down. We told him we would snag it with a big lure and pull it back up. And if we weren't successful in retrieving it that way, we would even swim down and get it! Mike was somewhat pacified but still didn't have a lot of confidence that we could recover the tackle box for him.

The next day, Wednesday, Bud and I cooked pancakes for breakfast – for everybody! Some of the younger kids who had never before camped out or cooked for themselves hadn't eaten much on Tuesday so they took full advantage of the "free" breakfast. After breakfast everybody was anxious to go fishing. The boys went off in pairs (the buddy system was in effect) to try their luck on the lake. Bud and I fished for much of the morning but in the afternoon we decided to see if we could hook Mike's tackle box.

By laying down on the platform and shielding our eyes from the sun we were able to see, down within the criss-crossed log foundation of the structure, Mike's tackle box. We could even make out one handle on the side of the box. We tied a 20-pound test fishing line onto a heavy spinner that had a large treble hook. Then we carefully lowered it toward the tackle box but it slipped off to the wrong side away from the handle. If the lure did drop to the side of the box where the handle was then the hook would slide off to the side or even right over the handle without snagging it. We tried over and over for more than an hour before giving up for the day. On Thursday we again tried, over and over, but again gave up.

When Friday came we were quite discouraged about the prospects of snagging the tackle box. It was now time for serious action. We had promised Mike that, if necessary, we would swim down to retrieve it. Now was the time to make good on our promise. There was a sandy beach on the north side of the lake about thirty yards from the outlet. We chose it as our base of operations for the salvage job.

All of the boys accompanied Bud and I as we approached the beach. They were eager to witness the spectacle of a man diving into the bone-chilling water. Bud started taking off his shirt even before we reached the beach and that was okay with me. Only one of us would have to do the swimming and we hadn't previously decided which of us it would be.

Bud stripped off his clothes and stood on the edge of the beach. The sun was shining and there was only a slight breeze so it was almost warm. But then Bud started to wade out into the lake and the water was so cold that by the time he was ankle-deep, his feet had turned blue. He hurriedly backed out and stood on the beach reviewing the situation again. There was only one thing to do. He waded out into the lake again and made it up to his knees before backing out again. Now his feet and his legs clear up to his knees were blue! The situation was really extreme now. There couldn't be any more backing out! He was going to wade out to above his knees and then dive into the water. With great resoluteness he prepared for the final plunge. He waded out past his ankles, above his knees, up to his thighs, and then – just as he was about to dive, the huge island of ice came drifting over from the middle of the lake and bumped against him! The slushy ice was a foot thick. He couldn't dive no matter how determined he was. So he backed out of the water and put his clothes back on.

We debated what to do next and decided to try the snagging exercise again. We returned to the outlet structure and, with the fishing line and the same large spinner, I once again carefully lowered the large treble hook down to the box. On the very first try it slid off the box to the side where the handle was. Then I very carefully raised the hook. It snagged the handle! We were highly elated by this success. But we weren't home free yet. The box still had to be raised up through the structure in which some of the criss-cross logs projected out farther than others. Was the handle solidly hooked? Would the handle of the box hold? Would the hook and line hold? We hardly dared to breathe as I pulled on the line and felt the weight of the box. It was solidly hooked. Slowly, slowly I raised the box from the bottom and up through the log structure. I cautiously moved it to one side or the other to avoid hitting it against any of the logs. And then it was within reach! Bud leaned over the platform and grabbed the box in both hands!

We had succeeded! Bud handed the box to Mike who was overjoyed – he took the box over to a level place on the bank and opened it. All of his Dad's lures and other gear were still there and in good shape. I don't know whether Mike's Dad ever found out what had happened with his tackle box or not.

I still look back and think that we were lucky that the ice drifted against the shore when it did. The results could have been disastrous for Bud if he had dived into the lake and swam down to the outlet. The exceedingly cold water, the depth of nineteen feet, and the suction of the water running out through the culvert all combined to create an extremely dangerous situation.

HIKING IN HIP BOOTS

Boy Scouts in the out-of-doors can be the source of great amusement and adventure. Another of our fishing trips to Kidney Lake was one of these. Over and over, year after year Bud and I would provide the boys with a list of things to bring on the trip. The boys were always tempted to wear gym shoes on the hike from Moon Lake to Kidney Lake. We always told them to wear hiking boots – no gym shoes. So it was with great dismay that we found, just as we were ready to start up the trail from Moon Lake, that Terry was wearing rubber hip boots! He didn't have any other shoes with him. We were sure that his feet would give out part way up the trail. However, there was no other choice but to let him try.

It had rained the evening before but the trail was dry or only slightly damp as we started out. We hiked along for about a mile and Terry was doing alright. Then we saw a tiny bit of snow in the shadow of the larger tree trunks. As we climbed higher there was more snow in the shadows. Then there was a tiny bit of slush on the trail. Not enough to bother us though. Terry was still keeping up and had no complaints. Then the slush was an inch deep and was a real nuisance. As we walked along our shoes would lift the slush up higher onto our pants. The slush kept increasing in depth as we climbed higher up the mountain. Terry was not just keeping up with us but he was moving along better than we were and he had absolutely no complaints! With a couple miles still to go the slush was almost six inches deep and all of us were thoroughly soaked and cold clear up to our knees. But not Terry! He was forging ahead and his feet and legs were as dry and comfortable as when he had started.

When we finally made it to Kidney Lake we weren't sure if we would ever get warm and dry again. Except for Terry whose feet and legs weren't chilled in the least and were totally dry! He was thoroughly enjoying his superiority over the rest of us. We had to sweep the slush away with pine boughs so that we could set our tents up and make fire places. There was sufficient dry "squaw" wood on the trees and we soon had fires going and warmed ourselves up. Luckily for us it didn't storm any more and the slush all melted the next day. Bud and I had to admit that Terry's hip boots were ideal for hiking up the trail – at least for that year!

LAUNCHED INTO SPACE BY A BROKEN ROPE

The annual Spring fishing trip of the Copperton Boy Scout Troop was scheduled for late June. Many of the boys who had participated in the previous fishing trips had graduated to the Explorer Scouts. This presented a problem. The Explorer leader position was temporarily vacant and the Explorers considered themselves to be too old and mature to camp with the Scouts. But they wanted to go on the fishing trip. In previous years they had mastered the skills of camping and caring for themselves. A compromise was reached whereby they would go on the trip under the direction of the Boy Scout leaders but would camp a hundred yards away and fend for themselves with only periodic inspections by the Scout leaders.

All preparations for the trip, the drive to Moon Lake, the hike up to Kidney Lake and camp setup went well. The Explorers were very anxious to show that they could manage themselves and needed no help or direction from Bud or me. There were a few anxious moments however, as when we stopped to visit them just as the grease in one of their frying pans caught fire and flames shot a foot into the air. The poor kid had intended to exhibit his cooking skills and show how independent he was. All in all though the Explorers did real well and were having a good time. But they were big boys and had a need to show us in some outstanding way just how big they were. They decided to do this by way of a challenge to a sporting event.

On the next to last day the Explorers issued a challenge to the Scouts for a tug-of-war. And that included the Scoutmaster. The site of the tug-of-war would be at the creek between the two camps. One team would be on each side of the creek. The Explorers intended to show how strong and tough they were by dragging the Scouts through the creek. There were only five Explorers but they were even bigger than me, the Scoutmaster. However, there were eight Scouts plus myself so we outnumbered them. The Explorers were confident that they would win. The Scouts were worried. Since the Explorers had made the challenge this meant the Scouts had the right to choose which side of the creek they would be on.

The Scouts chose the higher ground. There was a 3-foot bank at the Scouts' side of the creek and the ground gently rose upward from there. The Explorers would be on the lower ground on the other side of the creek. The ground was so low that some of it was almost swampy with damp soil and even some mud.

We had to piece together several lengths of rope and some of it was rather old and worn. The Scouts spaced themselves out along the rope and kicked footholds into the dry earth. Then, as a finale to our preparations, each boy made a loop in the rope so that he could put his hand through and hold onto the knot. We were ready! One of the youngest boys, Johnny, was first in line and I was right behind him. Johnny was very worried but I assured him that we were going to win.

Bud gave the signal and both sides started pulling with all their might! The rope was taut and quivering with the strain. Then, slowly, it became apparent that the Scouts were winning! Inch by inch the Explorers were being dragged toward the creek. The Explorers frantically heaved and jerked but they were inexorably drawn toward the creek and the muddy ground that was next to it. When the first Explorer got into the mud he started slipping and losing his balance. Then the second one got into the mud and the Explorers were threatened with an absolute rout.

But then there was a loud "Snap!" The Explorers all abruptly fell backward with the rope still in their hands. They had won! The Scouts had given up and let go of the rope! The Scouts (all except for Johnny) also fell backward and thought they had won. As I fell back I saw Johnny launched into the air toward the creek. Bud also witnessed Johnny's flight but none of the boys had seen it (except for Johnny and it took him a while to realize what was happening to him). The scene was one of absolute chaos. The Explorers, thinking that the Scouts were trying to deny them their victory, charged the Scouts' position with the intent of throwing each of them into the creek. The Scouts thought the Explorers were sore losers who were determined to salvage at least part of their pride after losing the tug-of-war. Soon there was a melee of wrestling and shouting boys. The Explorers were having their way and were dragging individual boys to their fate. Bud ran to the aid of the Scouts and with considerable shouting and forceful action finally stopped the battle.

Up to that point nobody but Bud, Johnny and I realized exactly what had happened. The rope had broken right behind Johnny! With his right hand and wrist through the loop he was effectively ensnared by the rope and couldn't let go. Since Johnny was the first in line on the rope, the very considerable force of the five Explorers, who were pulling with all their might, was suddenly acting upon only him. He was jerked into the air and, with his arms and legs outstretched, went flying into space and then splashed down into the creek! Johnny wasn't injured by his unintended feat. He is immortalized in Copperton Boy Scout history as the only Scout who was launched into space by a broken rope.

A NICOTINE FIT IN THE UINTAS

One particularly amusing incident (from my perspective at least) involved one of the boys' father. The scout troop was sponsored by the L.D.S. Church which practices the Word of Wisdom that frowns upon the drinking of alcoholic beverages and of smoking. All of the boys weren't LDS however and neither were their dads.

We had started a practice of taking the boys on a week-long fishing trip each spring to Kidney Lake in the Uinta Mountains above Moon Lake. All of the dads were invited to go with us but very few did. On this particular trip though, one of the dads (Carl, a nonmember who smoked heavily) accompanied his son. Carl helped us considerably by providing his station wagon for transportation. Bud, Carl and I, each with a vehicle loaded with camping supplies and boys, left Copperton early on Monday morning and drove to Moon Lake. Then we loaded all the gear onto horses that were provided by the wrangler at Moon Lake. He led the heavily laden pack string up the trail to Kidney Lake. The boys and we leaders carried only our fishing poles and a lunch as we walked the seven miles up the trail. Camp was set up alongside the creek that ran from Kidney Lake down into Island Lake. Each two boys shared a tent and a campfire on which they did their own cooking. Carl and his son shared a tent. Bud and I shared another tent. After setting up camp we even had some time to go fishing before going to bed for the night.

The next morning we were all busy cooking breakfast and getting ready to go fishing when four men on horses came riding up the trail. They stopped to talk with us for a few minutes. They were going up over the pass to the west past Tworoose and Rudolph Lakes and then farther up into the Squaw Basin. They said they would be coming back in three days. Carl noted that a couple of them had cigarettes in their shirt pockets. We watched them ride up the trail and then we went back to our cooking.

After breakfast we all went out fishing – some to the inlet to Kidney Lake, others to the sandy beach on the northeast side of the lake and some down to Island Lake. Later that afternoon I looked down toward our camp from Kidney Lake and saw Carl. It was evident, even from a quarter mile away, that something was wrong with him. He was extremely nervous and could hardly sit still. He would sit down and then abruptly jump back up again and then walk a short distance first in one direction and then in another. He couldn't stay in one spot for more than a minute or two. When we all got back to camp that evening the answer to Carl's problem became known. He had run out of cigarettes! He was having a nicotine fit! Nobody else in our camp had any cigarettes. Nobody else had come along the trail. There were only a few possible solutions to his problem -- such as following the horsemen to their destination and borrowing from them. Another was to walk back to Moon Lake to buy cigarettes. Or he could stay in camp by the trail with the hope that somebody else would come along in the next day or two and give him some. The last solution was to wait until the horsemen returned from their three-day trip. But it was already too late in the evening to do anything except to have supper and then go to bed.

Carl could hardly sleep that night. The next morning brought no relief. After carefully weighing his options he decided to wait by the trail for the horsemen to return. By doing this he could also catch anybody else who might happen along the trail before the horsemen returned. But that didn't happen. Nobody else came. He walked up the trail a couple hundred yards in the direction the horsemen had gone and saw that after passing through a thin strip of trees on the north side of Kidney Lake it came to a huge boulder about eight feet high. The trail went alongside the boulder and then across a small meadow before heading up the hill through thicker timber. Carl walked back along the trail and puttered around the camp. He sat down and got back up and walked a short distance and then sat down and got back up again. This was really tough on him! There was no relief all that day. He endured another difficult night. Morning finally came but again with no relief. He fussed and fidgeted, sat down and jumped back up, walked first one way and then another for the whole day. But again, nobody came along the trail and he was faced with another excruciatingly long night. The next morning Carl awoke early and walked up the trail to the big rock on the edge of the meadow. Then he came back to camp but not for long. He returned to the boulder, climbed up on top of it and sat down facing the far side of the meadow where he expected the horsemen to emerge. Carl sat perched atop the boulder like a huge buzzard watching for some prey. An hour went by. Still he waited. And then – he heard the sound of the horses' hooves as they approached through the timber.

The first horseman came out of the trees and Carl climbed down from the rock. He hurried toward the horseman and they met in the middle of the meadow. The man on the horse said "Howdy!". Carl didn't waste any time with preliminaries. He blurted out "Have you got any cigarettes?" The guy said "Yeah" and reached into his shirt pocket to pull out a pack. He handed a cigarette to Carl who immediately put it into his mouth and lit it with a match. The guy on the horse watched as Carl eagerly inhaled a big draft of smoke. Then he said "How long have you been out?" Carl, sounding like a shipwreck survivor who had been adrift for weeks in a shark-infested sea, said "Three days". The guy on the horse said "Wow that must have been tough!" Carl said "Yeah". By then the other three guys were sitting on their horses watching with sympathy as Carl sucked on the cigarette. One of them pulled out a full pack and gave it to Carl. They left him with enough cigarettes to last him for two more days. Then they rode off down the trail looking much like the successful rescuers of a shipwrecked sailor. And Carl looked like a shipwrecked sailor who had been saved after weeks adrift.

From "Bingham and Mines" Compilation by Gordon Bodily – Original Source Uncertain

The most productive and richest placer was that of Clays bar, from which $100,000.00 in gold had been taken out by 1868.

This bar consisted of a shaft which had to be abandoned due to excessive water encountered below a depth of 120 feet. Dan W. Heaston suggested that even though the shaft had to be abandoned, the old tailings might, if re-worked, produce a fair wage. The tailings were thrown into the sluice boxes and the result was wages ranging from $7.00 to $15.00 per day for the three men who owned the bar.

One day while Ben Clay was cleaning out the sluice boxes, throwing the larger cobbles high upon a dump, one piece rolled off the shovel. He picked it up and discovered it to be a gold nugget weighing eight ounces. This was said to be the largest nugget ever washed out in the state and was valued at $128.65.

LAST PHOTO AND REQUEST FOR INFORMATION

Figure 142 – Thousand Lb Mule Going Down 4-Foot Square Shaft In 1890s

Photo Courtesy David M. Pinney Through Gordon Bodily. This photo is just too good to leave out of this book even though it might not have been taken in Copperfield or even in Bingham Canyon. Edward G. Cupit, who was born in 1861, is standing in the foreground; he was probably in his 30s when this photo was taken. The name of the other participant is unknown. These two men are putting a thousand pound mule down a 4-foot square shaft that is 1300 feet deep. Both men appear to be quite satisfied with their handiwork. The onlookers seem to consider this an unusual spectacle that they just have to see. The mule is bound up in a cargo net and appears to be resigned to his fate since he can't do much about it anyhow.

Mules were used to pull cars of ore and waste in the underground mines and also on the surface. Most of the shafts were, however, larger than 4 feet square so the mules could stand on their feet in a larger cage as they were let down a shaft. Mules were sometimes kept underground for years without seeing the sunlight. They were fed and watered just as if they were in a barn and they were given preferential treatment over the miners. Mules were used in Copperfield to pull the empty ore cars back uphill on the Holden Tramway and other photos show mules on the surface at the Old Jordan and Galena mines in Galena Gulch.

Any additional information or comments regarding this photo or other contents of this book will be very welcome. The author intends to write another book about Bingham Canyon and is constantly seeking new photos and human interest stories related to the town and the mines.

THREE FIGHTERS NEEDED FOR TILT

New Fight Card Being Arranged for Local Fight Fans – Much Secrecy Being Used by Managers

Wanted: Three fighters who can stay twelve rounds with a real fighter. This is the sum and substance of a notice posted in the Copper King during the past week, and with the publication of the notice is an air of mystery that bodes something exciting for the local followers of the fistic arena.

Although very little can be learned at the Copper King, aside from the announcement, it is understood that some one in camp has a near champ who claims to be able to stop three local men in 12 rounds and will give $25.00 to the person who stays in the ring for four rounds, and $50.00 to anyone who can stay longer.

It is also announced on the blackboard of the popular billiard parlor that there will be four good preliminaries, which would be taken to mean that a real fistic card will be given, aside from the main event.

Who the manager is, is still unknown and who the said fighter is, who promises to mete out the sleep potion to three local fighters, is also an unknown quantity, and all the fans can do is sit back patiently and await further developments.

To date, one local fighter, Jack Carpenter, has signified his willingness to tangle with the unknown, and it has been intimated that the other two will be forthcoming before many days have passed.

Thus it is up to the fans to await the next move of the promoters, whoever they may be, and see what happens. Either way it looks as though Bingham Canyon will be treated to another keen fistic card.

The above article was printed in the November 21, 1924 edition of The Press-Bulletin, Bingham Canyon, Utah.

BIBLIOGRAPHY

1. Arrington, Leonard J. and Hansen, Gary B., *The Richest Hole on Earth*, Monograph Series, Vol. 11, No. 1, October 1963, Utah State University Press, Logan, Utah.
2. Bailey, Lynn R., *Old Reliable*, Westernlore Press, Tucson, Arizona, 1988.
3. Bingham Commercial Club, *Bingham: The Greatest Copper Camp In the World,* Bingham Canyon, Utah 1909.
4. Boutwell, J.M., Keith, Arthur, and Emmons, S.F. "*Economic Geology of the Bingham Mining District, Utah.* Professional Paper No. 38, United States Geological Survey. Washington, D.C.: Government Printing Office, 1905.
5. Butler, B.S., Loughlin, G.F., Heikes, V.C. et. al. *The Ore Deposits of Utah.* Professional Paper No. 111, United States Geological Survey. Washington, D.C.: Government Printing Office, 1920.
6. Carr, Stephen L. and Edwards, Robert W., *Utah Ghost Rails,* Western Epics, Salt Lake City, Utah 1989.
7. Crump, Scott, *Bingham Copper Mine - The Richest Hole On Earth*, Great Mountain West Supply, Salt Lake City, Utah 2005.
8. Crump, Scott, *The First 100 Years, A History of Jordan School District*, Jordan School District, Sandy, Utah 2005.
9. Dunn, Marion, *Bingham Canyon*, Publishers Press, Salt Lake City, Utah 1973.
10. Jackson, Arilla Bullock, *Copperfield*, Publishers Press, Salt Lake City, Utah 1984

PHILANTHROPISTS WILL BUY ELK

The proposition of hunting elk in this section suddenly came to a close this week, and many good sportsmen will be disappointed when they read this. Some weeks ago it was decided that in view of the fact that the males greatly outnumber the females in the local herd, and that they now fight a great deal, that it would be a good idea to kill off a few of the males and with the money secured purchase a number of does. So local sports got themselves in readiness for the big hunt, and already several little attempts have been made.

But this all came to an end this week when Jerome Bourgard approached the state deputy game warden, A. L. Heaston, and asked why it was necessary to kill off any of the males. Mr. Heaston informed him that the herd was not doing well on account of the fact that the males greatly outnumbered the females. To this Mr. Bourgard said that he would pay half the expense of getting a car load of does to add to the herd. Mr. Heaston is just the kind of a man to take up such a proposition, so without a moment's hesitation he told Mr. Bourgard that he would pay the other half.

A short while later Mr. Heaston called at this office and stated that the hunters might call off the sport, as a car load of does would be purchased to add to the herd. Mr. Bourgard and Mr. Heaston are both good men and the community will appreciate the steps they have taken.

The above article was printed in the February 15, 1918 edition of The Press-Bulletin, Bingham Canyon, Utah.

INDEX

BAN ON MASKS IS RAISED

At a meeting of the town board Wednesday night an order was passed dispensing with the compulsory wearing of masks. The report of the health officer showed conditions to be in good shape and the action was recommended by Mr. Schwartz of the State Health department.

Also the churches will be opened and the movies have started anew. Attendants at these public gatherings will be required to wear masks. There has been no change in the quarantine and persons leaving town must secure permits in order to be able to return.

Masks Have Been Effective

It is stated that the flu masks are doing much good aside from the prevention of contracting the flu. One party says that for a long time he had been troubled with a cough and that he has found great relief by wearing the mask, probably due to the fact that the dust of the air is prevented from reaching the throat or lungs.

There is a great deal of dust and smoke in the atmosphere here, due to the blasting at the surface mines on the Utah Copper hill. A great number of people here have been affected with a cough which is very noticeable by newcomers, and since the mask regulation went into effect this coughing has been greatly reduced. But in this connection only those who have suffered by reason of the cough have noticed it. The party of which we have reference wears his mask practically all the time. He works at one of the mines and wears the mask while at work because he has found relief in it from the cough.

There are some people here, like Mr. E. G. Locke, who ridicule the idea of the mask and say that what is needed is fresh air, as though the air would be contaminated by going through the mask but the fact is that the air is freed from the injurious dust by use of the mask. Furthermore the wearing of masks or something has practically placed the flu epidemic under control. One thing at least can be said of the mask and that is that it does no harm. It is a little inconvenience, but that is nothing compared with the flu.

The above article was printed in the December 13, 1918 edition of The Press-Bulletin, Bingham Canyon, Utah.

PURCHASE ORDER

Books can be purchased over the internet at BinghamCanyon.net or by mail from:

BinghamCanyon.net
c/o Eldon Bray
2186 McKinley Drive
Grand Junction, CO 81503

YOUR NAME AND ADDRESS

Name ____

Street ____

City ____

Zip ____________ **Phone #** ____________________

PAYMENT BY CHECK OR MONEY ORDER PLEASE

Signature ____

ITEM WANTED		QUANTITY	COST	TOTAL
Book COPPERFIELD REMEMBERED				
One or two copies	$24.95 each	________	X $24.95	________
Three to five copies	$21.95 each	________	X $21.95	________
Six to ten copies	$19.95 each	________	X $19.95	________
Sub Total				________
S&H $3/Order				$3.00
TOTAL ORDER		# ______		$________

OHIO COPPER CO. WILL PRECIPITATE

The Ohio Copper Co. have recently acquired more water rights in Bingham and will improve and extend their present precipitation plant in the Mascotte Tunnel. The company recently purchased a car load of clippings and small pieces of tin cans from a firm in Ogden which have proven more lucrative than the cans which had accumulated rust.

The above article was printed in the August 1, 1922 edition of The Press-Bulletin, Bingham Canyon, Utah.

OHIO COPPER'S GREATEST QUARTER

Figures recently submitted by General Manager Samuel Kellock of the Ohio Copper company to the directors show that the first quarter of 1924 was the most successful one in the history of the company. A total of 2,898,575 net pounds of copper were derived from the company's leaching in place operations.

Average per pound operating cost was 6.21 cents, which in itself is a record. Profit from operations for the quarter, according to Mr. Kellock's statement was $155,670.10.

During the quarter, a total of 153,252,000 gallons of water was pumped to the slopes of the company's property at Bingham and distributed. This water carried in solution 2,611,793 lbs of copper.

Of this total, 2,582,106 pounds were recovered and 79,687 lbs lost in the tails, making a recovery of 96.83 per cent for the quarter. Precipitates produced aggregated 3,847,660 pounds, of which 1,077,098 pounds was moisture. Precipitates carried 91.39 per cent copper.

Average price received for copper was 12.718 cents a pound which left a profit of 6.508 cents after deducting the operating cost of 6.21 cents.

The above article was printed in the April 26, 1924 edition of The Press-Bulletin, Bingham Canyon, Utah.